Praise for *No Beast So Fierce*

"A vivid portrait. . . . Huckelbridge writes with authority and clarity, deftly weaving strands of economics, sociology and history, explaining how changes to a land and its people upset natural systems that had held for millennia. . . . *No Beast So Fierce* excels as an intelligent social history and a gripping tale of life and death in the Himalayan foothills."
—*Star Tribune* (Minneapolis)

"Thrilling. . . . Fascinating. . . . Exciting." —*Wall Street Journal*

"Gripping. . . . From 1900 to 1907, a female Bengal tiger (*Panthera tigris tigris*) killed hundreds of villagers in northern India and Nepal. This compelling account hinges on that grisly story, but digs deep into causation." —*Nature*

"I had a feeling this book would hook me from the get-go. I was right. *No Beast So Fierce* is much more than a cautionary tale of the Man-Eater of Champawat, a Royal Bengal tiger responsible for hundreds of deaths in Nepal and India, or of Edward James Corbett, the legendary hunter who shot and killed the big cat in 1907. Dane Huckelbridge's remarkable narrative also reveals the circumstances that cause tigers to stalk human prey as well as Corbett's transformation into a conservationist and ardent champion for protecting the animals he once hunted."
—Michael Wallis, author of *The Best Land Under Heaven: The Donner Party in the Age of Manifest Destiny*

"The fascinating tale of the Champawat Tiger, [and] also the story of the forces that created her. . . . This multilayered approach to what is, at heart, the account of [Jim] Corbett's long-term hunt for the famous man-eater elevates Huckelbridge's book above the sensational 'true tale' to stand as a superb work of natural history." —*Booklist* (starred review)

"A great tale and study of man versus beast, or, rather, beast versus man. The seminal battle between Jim Corbett and the Champawat Tiger stands as an epic encounter of the ages. Dane Huckelbridge's *No Beast So Fierce* will make you rethink your position in God's universe—and on the food chain."
—Jim DeFelice, #1 bestselling coauthor of *American Sniper* and author of *West Like Lightning: The Brief, Legendary Ride of the Pony Express*

"Huckelbridge showcases his storytelling skills effectively in this suspenseful look at 'the most prolific serial killer . . . the world has ever seen.' . . . A gripping page-turner that also conveys broader lessons about humanity's relationship with nature." —*Publishers Weekly*

"Huckelbridge details the surprisingly methodical and incredibly bloody machinations of a single Bengal 'tigress.' Between 1900 to 1907, the Champawat man-eater stalked humans living in the villages of southern Nepal and, because tigers know no borders, eventually northern India. Along her route, she killed 435 people, making her perhaps the most murderous non-human animal in recorded history."

—*Popular Science*

"*No Beast So Fierce* is an exciting read well worth your attention."

—Asian Review of Books

"At the heart of *No Beast So Fierce* is a simple and terrifying story: In the early 20th century, a tiger killed and ate more than 400 people in Nepal and northern India before being shot by legendary hunter Jim Corbett in 1907. Rather than just describe this harrowing tale, though, author Dane Huckelbridge seeks to explain how such a prolific man-eating tiger came to be, taking readers on a fascinating journey through the natural history of a tiger and the political history of Nepal and northern India. . . . Satisfy[ing]." —*Science News*

"Absorbing. . . . An awesome recapitulation of the fearsome events surrounding this notorious killer and the hunter who finally trails and shoots the dreaded animal. . . . Thrilling, bonechilling."

—SanFranciscoBookReview.com

"[A] gripping tale. . . . [A] mesmerizing man-beast encounter that Huckelbridge narrates quite brilliantly. . . . A fitting tribute to Jim Corbett and the Bengal tiger." —*Khaas Baat*

"One of the great adventure stories of the 20th century. . . . A real-life thriller." —*Princeton Alumni Weekly*

"The story of the Champawat Tiger and his pursuit by Jim Corbett is told in this tremendous book. . . . *No Beast So Fierce* reads like part history, part thriller. . . . The book often reads like a detective novel. . . . Huckelbridge tells this story well and you walk away from this book feeling rewarded with knowledge as well as entertained."

—HuntingLife.com

"The search for the man-killer known as the Champawat Tiger provides a suspenseful narrative. . . . The reader is likewise introduced to the history of Nepal and British colonialism in India and the man-made environmental changes which inevitably forced animals content with their natural jungle superiority to develop appetites for human flesh. . . . Huckelbridge continues to demonstrate his versatility and skill as a writer with this book." —*Journal Star* (Lincoln)

NO
BEAST
SO
FIERCE

NO
BEAST
SO
FIERCE

The Terrifying True Story of the
Champawat Tiger, the Deadliest
Animal in History

DANE HUCKELBRIDGE

WILLIAM MORROW
An Imprint of HarperCollinsPublishers

HarperCollins books may be purchased for educational, business, or sales promotional use. For information, please email the Special Markets Department at SPsales@harpercollins.com.

A hardcover edition of this book was published in 2019 by William Morrow, an imprint of HarperCollins Publishers.

FIRST WILLIAM MORROW PAPERBACK EDITION PUBLISHED 2020.

Designed by William Ruoto

The Library of Congress has catalogued a previous edition as follows:

Names: Huckelbridge, Dane, author.
Title: No beast so fierce : the terrifying true story of the Champawat Tiger, the deadliest animal in history / Dane Huckelbridge.
Description: New York, NY : William Morrow, [2019] | Includes bibliographical references and index.
Identifiers: LCCN 2018024761| ISBN 9780062678843 (hardcover) | ISBN 9780062678867 (trade paperback)
Subjects: LCSH: Tiger hunting—India—Anecdotes. | Tiger hunting—Nepal—Anecdotes. | Tiger—India—Anecdotes. | Tiger—Nepal—Anecdotes. | Corbett, Jim, 1875-1955.
Classification: LCC SK305.T5 H83 2019 | DDC 799.2/7756—dc23
LC record available at https://lccn.loc.gov/2018024761

ISBN 978-0-06-267886-7 (pbk.)

20 21 22 23 24 LSC 10 9 8 7 6 5 4 3 2

No beast so fierce but knows some touch of pity. But I know none, and therefore am no beast.

—William Shakespeare, *Richard III*

Do not blame God for having created the tiger, but thank Him for not having given it wings.

—Indian proverb

CONTENTS

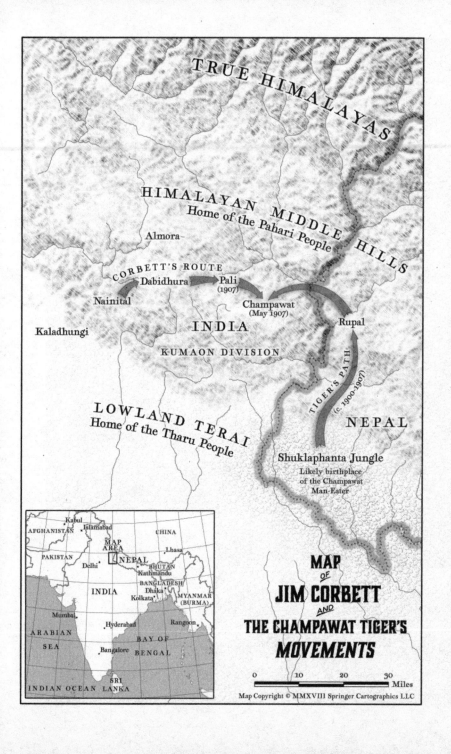

TRUE HIMALAYAS

HIMALAYAN MIDDLE HILLS
Home of the Pahari People

Almora

CORBETT'S ROUTE
Dabidhura → Pali
(1907)

Nainital

Champawat
(May 1907)

Rupal

Kaladhungi

INDIA

KUMAON DIVISION

TIGER'S PATH
(c. 1900–1907)

LOWLAND TERAI
Home of the Tharu People

NEPAL

Shuklaphanta Jungle
Likely birthplace
of the Champawat
Man-Eater

Kabul
AFGHANISTAN   Islamabad   CHINA
              MAP
PAKISTAN   AREA   NEPAL   Lhasa
         Delhi   BHUTAN
                    Kathmandu
                    BANGLADESH
                 Dhaka
              Kolkata   MYANMAR
                         (BURMA)
INDIA

Mumbai
              Hyderabad
                              Rangoon
ARABIAN
SEA   Bangalore   BAY OF
                           BENGAL
INDIAN OCEAN   SRI
                LANKA

MAP
OF
JIM CORBETT
AND
THE CHAMPAWAT TIGER'S
MOVEMENTS

0   10   20   30
                              Miles

Map Copyright © MMXVIII Springer Cartographics LLC

PROLOGUE

|||||||||||||||||||||

We do not know the year. Nor does history record the poacher's name. But around the turn of the twentieth century, somewhere in the *terai* near the Kanchanpur District of western Nepal, a man made a terrible mistake.

He attempted to kill a Bengal tiger.

We can imagine him to be a young man—that seems all but certain. For the local Tharu people are well acquainted with tigers, and only a youthful and inexperienced hunter would be so careless. After all, a tiger hunt among the Tharu is a solemn affair, to be initiated with a puja sacrifice of roosters and goats, as a show of respect to the forest deity Ban Dhevi. It is an act of profound spiritual and earthly significance, one that risks angering gods and kings alike. If such a decision is to be even considered, it must be blessed by a *gurau* with a sacred glass of *rakshi,* and sanctified by the wearing of holy red ribbons.

But change is coming, even to this remote province. Like others of his generation, this brash young man likely may have tasted the British gin and cigarettes that come smuggled across the border from India, and seen the Western suits and cravats one can purchase beyond the Sharda River, and he has no time for rice liquor or garlands made of ribbons. He does not see the tiger as a divine spirit, a lord of the forest, a custodian of the natural world, maintaining the balance of all things. To him, a tiger is a sack of gold and nothing more: money for clearing land, funds to buy a water buffalo and start a farm of his own. The young man bristles at the thought of eking out a living from the forest like his parents, of dwelling in a mud-walled house thatched with elephant grass. No, that is simply not for him.

So, we may imagine, he sets out from his village, a decrepit old muzzle-loader slung over his shoulder, an oblivious goat hobbling along in tow. He follows a path of packed earth, skirting the edge of the mustard and lentil fields, tracing the dry bed of a meandering nullah, until he at last reaches the sal trees where the true jungle begins. He has built a small machan—a tiger-hunting stand—near a clearing where he has seen fresh tiger tracks in the mud, and after tying the goat to a peg in the earth, he mounts his machan and does his best to get comfortable.

The heat of the afternoon mounts, and the goat flicks its ears lazily, and the odd croak of a mating florican is the only sound to be heard. The young man wipes the sweat from his brow and scratches a mosquito bite, his initial excitement turning slowly to boredom, and then at last to irritation.

The shadows lengthen, dusk approaches, and still the scrawny goat stands tethered and unmolested. The young man begins to doubt that the tiger will come at all. Perhaps the old men in the village were right, perhaps it was foolish to even consider coming into the forest without—

And then it happens. It arrives with a grace and a force unlike any the young man has ever seen. An attack appalling in its power and mesmerizing in its beauty, as if the dappled patterns of the forest floor themselves have come alive and engulfed the poor creature. A liquid blur of tawny stripes, then a mound of working muscle. The goat has time to neither move nor bleat—one moment it is alive, and the next, it is not. Its neck is snapped in an increment of time too small to measure.

The young man's purpose is suddenly called into question. The notion of shooting the tiger before him feels impossibly bold, as if he were not killing a mere animal, but assassinating a king. Its body appears enormous, even from the safety of his machan. Its eyes are closer to those of a man than a pig or a deer, or any other creature he has encountered. And as if to further sour his conviction, two cubs appear, bounding almost playfully from the trees behind it. This is not just a tiger—*she* is a mother.

But for all his fear, the idea of returning home with nothing but the frayed goat tether unsettles him even more. No, he has made his decision. It must be done. And with the mother down, it will be easy to finish off the cubs as well. That's two more tigers than he had bargained for. He takes the old muzzle-loader in his trembling hands, raises its battered stock to his shoulder, gets the tiger in his sights, and takes one final breath before pulling the trigger.

But that is enough. The rustle of his movements, however faint, are not missed by the tiger's spotted ears. It drops the goat and raises its head in alarm, as a thunderclap bursts from the trees—a red sting of pain lashes at its jaw. The tiger rears back, as if to attack the air, only to find that its bite feels loose and unhinged. The taste of its own blood filling its throat, the tiger turns and streaks back into the forest, into the thick underbrush from whence it came, its two toy-sized cubs hesitating for a moment before bouncing along obediently behind.

The young man reloads his gun and springs down from his stand, racing to see if his bullet struck home. He notices the trampled earth beside the pathetic carcass of the goat, and next to it, a spattering of blood and two broken teeth—tiger teeth. The young man realizes his shot was poor and the tiger merely wounded, a fact that is confirmed moments later by a roar that seems to rend the very fabric of the air. He has heard tigers before, their low moaning from a distance, but this is different. He has never experienced anything like this. He feels the roar as much as hears it, in the pit of his stomach and the hollows of his chest. It is the purest distillation of rage he has ever known.

Darkness is coming. The idea of going blind into the bush to confront the enraged tiger is beyond comprehension to the young man, a nightmare he can't even begin to consider. No, it would be suicide—courting death in its most primordial form. And so, still sick with adrenaline, he slings his antique gun over his shoulder and turns back to the village on weakened knees, first walking, then running, casting harried glances over his shoulder the whole way, covering his ears to stifle the roars. And while there is no way for this young man to know the full implications of what he has done, the terror he has unleashed, the lives he will have indirectly ruined, he must surely have an inkling, with those roars reverberating through the still air and damp leaves of the sal trees, that in the pulling of that trigger, he has created a monster.

NO
BEAST
SO
FIERCE

UNLIKELY HUNTERS

In the first decade of the twentieth century, the most prolific serial killer of human life the world has ever seen stalked the foothills of the Himalayas. A serial killer that was not merely content to kidnap victims at night and dismember their bodies, but also insisted on eating their flesh. A serial killer that, for the better part of ten years, eluded police, bounty hunters, assassins, and even an entire regiment of Nepalese Gurkhas.

A serial killer that happened to be a Royal Bengal tiger.

Specifically, a tiger known as the Man-Eater of Champawat. Far more than an apex predator that occasionally included humans in its diet, it was an animal that—for reasons that wouldn't become apparent until its killing spree was over—explicitly regarded our species as a primary source of food. And to that end, this brazen *Panthera tigris tigris* hunted Homo sapiens on a regular basis across the rugged borderlands of Nepal and India in the early 1900s with shocking impunity and an almost supernatural efficacy. In the end, its reported tally added up to 436 human souls—more, some believe, than any other individual killer, man or animal, before or since.

Despite its unusual appetites and hunting prowess, however,

surprisingly little has been written about the Champawat. And when the odd mention of the tiger does crop up, it is more often than not as a curious footnote to a broader article on human–tiger conflict, or as a gory bit of trivia from *The Guinness Book of World Records*. The fact that a single tiger was able to take such an immense human toll over such a long period of time is rarely presented as a subject worthy of historical scrutiny or academic study. It seems like a good story, and nothing more.

And admittedly, it *is* a fine story, and it is tempting to present it simply as such. It is universal in its appeal and almost literary in its Beowulfian dimensions: a man-eating creature that terrorizes the countryside, repeatedly evading capture, until a hero appears who is brave enough to track it straight to its lair. It is a timeless campfire tale, simple and hair-raising in the way all such yarns must be. Who wouldn't want to hear a story like that? One that speaks to the most primal and deeply ingrained of all human fears?

But there is another story to be told here as well, and while certainly hair-raising, it is anything but simple. The events that transpired in the forests and valleys of the Himalayan foothills in the first decade of the twentieth century were not a series of bizarre aberrations. They were in fact the inevitable result of the tremendous cultural and ecological conflicts that were shaking the region—indeed, the world—at that time, affecting man and animal alike in unlikely ways, and throwing age-old systems chaotically out of whack. Far from some pulp fiction tale of man versus nature or good versus evil, the story of the Champawat is richer and much more complex, with protagonists at odds with even themselves.

Beginning, of course, with the actual tiger. Bengal tigers do not under normal circumstances kill or eat humans. They are by nature semi-nocturnal, deep-forest predators with a seemingly ingrained fear of all things bipedal; they are animals that will generally change direction at the first sign of a human rather than seek an aggressive confrontation. Yet at the turn of the twentieth century, a change so profound and upsetting to the natural order was occurring in Nepal and India as to cause one such tiger to not only lose its inborn fear of humans altogether, but to begin hunting them in their homes on an all but weekly basis—a tragedy for the more than four hundred individuals who would eventually fall victim to its teeth and claws. This tiger ceased to behave like a tiger at all, in important respects, and transformed into a new kind of creature all but unknown in the hills of northern India's Kumaon district, prowling around villages and stalking men and women in broad daylight.

Then there is Jim Corbett, the now-legendary hunter who was finally commissioned by the British government to end the Champawat Tiger's reign. To many, even in present-day India, he is nothing short of a secular saint, a brave and selfless figure who risked life and limb to defend poor villagers when no one else would. To others, particularly academics engaged with post-colonial ecologies, he is just another perpetrator of the Eurocentric paternalism that defined the colonial experience. Each is a fair judgment. The whole truth, however, is far more nuanced, as one would expect when it comes to a deeply conflicted man whose life spanned eras, generations, and eventually even empires. Jim Corbett was a prolific sportsman who, upon achieving fame, hobnobbed with aristocrats and used tiger hunts to curry their favor. But he was also a tireless advocate for wild tigers and devoted the latter part of his life to their conservation—as evidenced by the sprawling and magnificent national park in India that bears his name to this day. Yes, he did come to enjoy the trappings and privileges of the English *sahib*, servants and sport shooting and social clubs included. But as the domiciled son of an Irish postmaster, foreign-born and considered socially inferior, he was also keenly aware of what it meant to be colonized—by the very people he enabled and admired. And he did love India, above all its people, even while playing an unwitting part in the nation's subjugation.

Which brings us, inevitably, to colonialism itself—a topic far too broad and multifaceted for any single book, let alone one that's concerned primarily with man-eating tigers. Yet it is colonialism, undeniably, and the onslaught of environmental destruction that it almost universally heralds, that served as the primary catalyst in the creation of our man-eater. It may have been a poacher's bullet in Nepal that first turned the Champawat Tiger upon our kind, but it was a full century of disastrous ecological mismanagement in the Indian subcontinent that drove it out of the wild forests and grasslands it should have called home, and allowed it to become the prodigious killer that it was. What becomes clear upon closer historical examination is that the Champawat was not an incident of nature gone awry—it was in fact a *man-made* disaster. From Valmik Thapar to Jim Corbett himself, any tiger wallah could tell you the various factors that can turn a normal tiger into a man-eater: a disabling wound or infirmity, a loss of prey species, or a degradation of natural habitat. In the case of the Champawat, however, we find not just one but all *three* of these factors to be irrefutably present. Essentially, by the late nineteenth century, the British in the United Provinces of northern India and their Rana dynasty counterparts in western Nepal had created, through a combination of irresponsible forestry tactics, agricultural policies, and hunting practices, the ideal conditions for an ecological catastrophe. And it was the sort of catastrophe we can still find whiffs of today, be it in the recent spate of shark attacks in Réunion Island, the rise of human–wolf conflict on the outskirts of Yellowstone, or even the man-eating tigers that continue to appear in places like the Sundarbans forest of India or Nepal's Chitwan National Park. In the modern day, we have at last, thankfully, come to realize the importance of apex predators in maintaining the health of our ecosystems—but we're still negotiating, somewhat painfully, how best to live alongside them. And that's to say nothing of the far more sweeping problems posed by global warming and mass extinction, exigencies that have arisen from very much the same amalgamation of economic mismanagement and environmental destruction. Apex predators are generally considered bellwethers of the overall health of the environment, and at present, with carbon emissions on the rise and natural habitats diminishing, the outlook for both feels disarmingly uncertain.

Which is why this particular story of environmental conflict is not only relevant, but urgent and necessary. At its core, Jim Corbett's quest to rid the valleys of Kumaon of the Champawat Tiger is dramatic and straightforward, but the tensions that underscore it contain the resonance of much larger and more grievous issues. Yes, it is a timeless tale of cunning and courage, but also a lesson, still very much pertinent today, about how deforestation, industrialization, and colonization can upset the fragile balance of cultures and ecosystems alike, creating unseen pressures that, at a certain point, *must* find their release.

Sometimes even in the form of a man-eating tiger.

PART I

NEPAL

||||||||||||||||||||

CHAPTER 1

THE FULL MEASURE OF A TIGER

Where does one begin? With a story whose true telling demands centuries, if not millennia, and whose roots and tendrils snake into such far-flung realms as colonial British policies, Indian cosmologies, and the rise and fall of Nepalese dynasties, where is the starting point? Yes, one could commence with the royal decrees that compelled Vasco da Gama to sail for the East Indies, or the palace intrigues that put Jung Bahadur in the highest echelons of Himalayan power. But the matter at hand is something much more primal—elemental, even. Something that's shaped our psyches and permeated our mythologies since time immemorial, and that speaks directly to the most profound of our fears. *To be eaten by a monster.* To be hunted, to be consumed, by a creature whose innate predatory gifts are infinitely superior to our own. To be ripped apart and summarily devoured. And with this truth in mind, the answer becomes even simpler. In fact, its golden eyes are staring us right in the face: *the tiger.* That is where the story begins.

"The normal tiger," writes Charles McDougal, a naturalist and tiger expert who spent much of his life studying the big cats in Nepal, "exhibits a deep-rooted aversion to man, with whom he avoids contact." This is a fact corroborated time and time again by biologists, park rangers, and hunters alike, all of whom can attest firsthand to just how shy and elusive wild tigers actually are. One can spend a lifetime in tiger country without ever laying eyes upon an actual tiger, with the occasional pugmark or ungulate skull the only hint at their phantomlike presence. Even for modern-day Tharu who live alongside reserves with dense predator populations, it's fairly uncommon to see a tiger. Sanjaya, who served as my host and guide in Chitwan while I was conducting research for this book, grew up fishing and foraging in local forests, and in all those years, he had spied a tiger just once. No, the *normal* tiger has little interest in our kind, and even less in challenging us to a fight. With hunting, mating, and fending off territorial rivals taking up most of its time, the normal tiger has more important things to worry about; we barely rate a passing glance. We are a nuisance to be avoided, and nothing more.

However, for the *abnormal* tiger—that is to say, the tiger that has shed for whatever reason its deep-seated aversion to all upright apes—there are essentially two ways it will kill a member of our species.

The first category of attack is a defense mechanism, a means of protection, and it is employed only when a tiger sees a human as a threat to its safety or that of its cubs. When a mother tiger is surprised in a forest, or when a wounded tiger is cornered by a hunter, its instincts for self-preservation kick in and the claws come out. This tiger will often roar, come bounding in a series of terrifyingly fast leaps, and commence beating its human target head-on with its front paws, with enough power in most cases to smash the skull after the first strike or two. And from there, it only gets worse— according to Russian tiger specialist Nikolai Baikov, once the offending human is on the forest floor, "the tiger digs its claws as deeply as possible into the head or body, trying to rip off the clothing. It can open up the spine or the chest with a single whack." This is strictly a combative behavior, the inverse of predation (although defensive attacks do sometimes result in consumption as well). It manifests itself when the tiger senses imminent danger, and for that reason, calls upon its considerable resources to save its own skin—figuratively, and, given the price a tiger pelt can fetch on most black markets, literally as well.

And the results of this behavior, as the rare individual who is both unfortunate enough to encounter it yet still fortunate enough to survive it can tell you, are understandably horrific. There exists a video—and a quick Internet search will readily reveal it—of one such attack that occurred in Kaziranga National Park in northeastern India in 2004. Filmed from atop an elephant, it shows a group of park rangers tracking a problem tiger that had roamed beyond the boundaries of the reserve and begun killing cattle—almost certainly as a result of diminished habitat and limited natural prey. Armed with tranquilizer guns, their intent was not to harm the tiger, but rather to capture it before angry farmers did, and return it safely to its home in the park. But alas, the four-hundred-pound cat was not privy to this plan. Although grainy, and shot with an unsteady hand, the film makes the terrific competence with which a tiger can protect itself abundantly clear. With astounding speed and athleticism, the roaring tiger materializes from the high grass as if out of nowhere, leaps over the elephant's head with claws at the ready, and with merely a single glancing blow, manages to shred the poor elephant driver's left hand to bloody ribbons before making its getaway. And this happened to a group of heavily armed men mounted on towering pachyderms—one can imagine what such a tiger could have done to a single individual alone in the forest. They would have been dead before they had time to squeeze off a shot, a fact supported by one lethal Amur tiger attack recorded in 1994 in the Russian Far East—the local hunter's gun was found still cocked and unfired, right beside his mittens, while his ravaged remains were discovered in a stand of spruce trees one hundred feet away.

There is, however, a second means of attack that the tiger employs when it regards something not as a threat, but as a potential food source—one that relies less on claws than it does upon teeth. Specifically, a set of three- to four-inch canine teeth, the largest of any living felid (yes, saber-toothed tigers are excluded), designed to sever spinal cords, lacerate tracheas, and bore holes in skulls that go straight to the brain. And it makes sense a tiger would have such sizable canines given their usual choice of prey: large-bodied ungulates like water buffalo, deer, and wild boar. Two of the Bengal tiger's preferred prey species—the sambar deer and the gaur bison—can weigh as much as a thousand pounds and three thousand pounds, respectively, which gives some idea of why the tiger's oversized set of fangs are so crucial to its survival. They are the most important tools at its disposal for bringing down some of the most powerful horned animals in the world. To crush the muscle-bound throat of a one-ton wild forest buffalo is no easy task, but it is one for which the tiger is purpose-made.

The tiger's evolutionary history, however, begins not with a saber-toothed gargantuan eviscerating lumbering mastodons, but with a diminutive weasel-like creature scampering among the tree branches. With miacids, more specifically, primitive carnivores that inhabited the forests of Europe and Asia some 62 million years ago. Bushy-tailed and short-legged, these prehistoric scamps lived primarily on an uninspiring diet of insects. Their bug-feast would apparently continue for another 40 million years, until the fickle tenets of evolutionary biology put a fork in the road—some miacids evolved into canids, which today include dogs, wolves, foxes, and the like, while a second group, over the eons that followed, would turn into felids, or cats. Initially, there were three subgroups of felids: those that could be categorized as *Pantherinae, Felinae,* and *Machairodontinae,* with the third, although today extinct, including saber-toothed *Smilodons* that were indeed capable of ripping the guts out of woolly mammoths, thanks to their foot-long fangs and thousand-pound bodies.

Tigers, however, arose from the first group. Unlike leopards and lions, which both came to Asia via Africa (there are still leopards in India today, as well as lions, although only very small populations survive in the forests of Gir National Park), tigers are truly Asian in origin, first appearing some 2 million years ago in what is today Siberia and northern China. From this striped ancestor, nine subspecies would emerge, to spread and propagate across the continent, of which six still survive today, albeit precariously.

The Bali tiger, the Javan tiger, and the Caspian tiger all went extinct in the twentieth century, due to the usual culprits of habitat destruction and over-hunting. The first vanished from the face of the earth in the early nineteen hundreds, although the latter two subspecies seemed to have hung on at least until the 1970s. The extermination of the Caspian tiger is especially unsettling given the sheer size of its range—the large cats once roamed from the mountains of Iran and Turkey all the way east to Russia and China.

Of the tiger subspecies that still exist today, the Amur tiger—also known as the Siberian tiger—has stayed closest to its ancestral homeland in the Russian taiga, and continues to prowl the boreal forests of the region in search of prey. This usually means boar and deer, although at least one radio-collared tiger studied by the Wild-life Conservation Society, or WCS, was recorded as feasting primarily on bears. It seems the Amur not only has a preferred method of killing bears, involving a yank to the chin accompanied by a bite to the spine, but that it is also somewhat finicky, preferring to dine on the fatty parts of the bear's hams and groin. With a thick coat, high fat reserves, and pale coloration, the Amur tiger is well suited to the wintry landscape of the Russian Far East. It is generally considered the largest of all the tiger subspecies, with historical records showing weights of up to seven hundred pounds, although a modern comparison of dimensions reveals that Bengal tigers in northern India and Nepal today are actually larger on average than their post-Soviet relatives farther east. Research hasn't effectively concluded why Amur tigers are physically smaller today than in centuries past, although the removal from the gene pool of large "trophy" tigers could well be a factor. The Amur tiger's current wild population numbers only in the hundreds, confined to a few pockets of eastern Russia and the borderlands of China.

While the northernmost realms of East Asia are prowled by what little remains of the Amur tiger population, the warmer climes to the south claim their own small subgroups of *Panthera tigris*, in the form of the Indochinese tiger, the Malayan tiger, the Sumatran tiger, and the South China tiger. All of their populations are atrociously small, with the South China tiger bearing the ignominious distinction of being categorized, at least recently, as one of the ten most endangered animals in the world, and quite possibly extinct in the wild. The future of all tigers is precarious at best, but in the case of these lithe jungle dwellers, considerably smaller on average than their brethren to the north, it is even more so.

Farther west, however, in the forests of India, Nepal, Bhutan, and Bangladesh, there is another tiger yet, and although endangered, it has miraculously managed to maintain numbers that border on the thousands. It is the tiger of Mughal emperors and maharajas, of Rudyard Kipling and William Blake. The tiger that Durga, the Hindu mother goddess, rode into battle to vanquish demons, and that the rebellious Tipu Sultan—also known as the Tiger of Mysore—chose as his standard. It is the tiger that yanked British generals from their howdahs atop elephants, that turned entire villages in Bhiwapur into ghost towns, and that is responsible for the vast majority of the million people believed to have been killed by tigers over the last four centuries. It is the tiger of nursery rhymes, the tiger of nightmares, the tiger our imagination conjures when the word itself is spoken. It is identified by scientists, rather prosaically if not redundantly, as *Panthera tigris tigris,* but it is known to all as the Bengal tiger.

\|\|\|\|\|\|\|\|\|\|\|\|\|\|\|\|\|\|\|\|\|\|\|\|\|\|\|\|\|\|\|\|\|\|

There is no shortage of shades or strokes one can employ when it comes to painting a portrait of the Bengal tiger, but to begin: Bengal tigers are *big.* While females tend to max out close to 400 pounds, adult males regularly achieve body weights in excess of 500 pounds, and some exceptionally large individuals have been documented at weights of over 700 pounds. Royal Bengal tigers, the subset that lives in the sub-Himalayan jungle belt known as the *terai,* tend to be even bigger. One extraordinary specimen—reportedly also a man-eater, at least until David Hasinger shot it in 1967—weighed in at 857 pounds, measured over 11 feet long, and left paw prints "as large as dinner plates." For its last supper, it managed to drag not only a live water buffalo into the forest, but also the eighty-pound rock to which it was tethered. The humongous tiger's man-and-water-buffalo-eating days may have ended shortly thereafter, but it still prowls today—in the Smithsonian Institution's National Museum of Natural History, in fact, where it is on permanent display in the Hall of Mammals.

Second: Bengal tigers are *fast*. In short sprints, they can achieve forty miles per hour, which is almost three times as fast as the average human being, and roughly equivalent to the top speed of a Thoroughbred racehorse. In other words, it is futile to try to outrun a dedicated tiger. And when it comes to their leaping prowess, there are plenty of examples of tigers clearing tremendous hurdles to get their claws on a target. In an incident recorded in Nepal in 1974, a startled tigress protecting her cubs had little trouble mauling a researcher hiding fifteen feet above her in a tree. The aforementioned tiger from Kaziranga National Park managed in 2004 to take three fingers off that unfortunate elephant driver's hand—a hand that appears to be at least twelve feet off the ground—with barely a running start. And on Christmas Day, 2007, a tiger (Amur, not Bengal, although their abilities are comparable) escaped the ostensibly inescapable barriers of its open-air enclosure at the San Francisco Zoo, for the sole purpose of going after a trio of young men who had provoked its ire. Accounts vary as to what caused the attack—the zoo accused the victims, all of whom had alcohol and marijuana in their systems, of taunting and harassing the animal, something the two survivors vigorously denied. What is certain, however, is that the enraged tiger got across a thirty-three-foot dry moat, cleared a nearly thirteen-foot protective wall, and emerged snarling from the pit like the wrath of God. Police arrived in time to save two of the young men from almost certain death—

the third, who received the brunt of the initial attack, was not so lucky—but stopping the crazed tiger proved anything but easy. One officer fired three .40-caliber-pistol rounds into the charging cat's head and chest, and that only seemed to anger it further. It wasn't until a second officer put a fourth bullet in the tiger's skull at point-blank range that it finally ceased its attack and fell to the ground. A more nightmarish scenario is difficult to imagine, but it is at least worth mentioning—this was only a *captive* tiger. Experienced wild tigers, accustomed to bringing down big game and fighting off territorial rivals, are generally much more athletic *and* aggressive when their hunting or defensive instincts kick in. A tiger in its natural habitat—alert, attuned, muscles rippling beneath its tawny striped hide—is another creature entirely from the languid, yawning pets of Siegfried & Roy. As lethal as this urbanized West Coast zoo tiger proved to be, it was a flabby house cat compared to its country cousins, ripping apart wolves and chasing down bears in ancient forests across the sea.

Third: Bengal tigers are *strong*. A tiger's jaw is capable of exerting around a thousand pounds of pressure per square inch—the strongest bite of any cat. That's four times as powerful as the bite of the most menacing pit bull, and considerably stronger than that of a great white shark. Even Kodiak bears, which can weigh as much as 1,500 pounds, can't keep up. The bite of a tiger can shred muscle and tendon like butter and crunch bones like we might a stale pretzel stick. And if their bite is terrifying, a swipe from their retractable claws is just as bad if not worse. A single blow from a Bengal tiger's paw can crack the skull and break the neck of an Indian bison, and can decapitate a human. Aggressive tigers have been known to rip the bumpers off cars, tear outhouses to splinters, and burst through the walls of houses in search of food. They can drag a one-ton buffalo across a forest floor with ease, and are capable of carrying an adult chital deer by the neck as effortlessly as a mother cat does a kitten. It's less apparent on an Amur tiger, with its heavy fur and fat reserves, but on a Bengal tiger, the musculature is unmistakable—this is the middle linebacker of the animal world, the perfect melding of power and speed.

And last: Bengal tigers are *smart*. Predation of almost any kind requires intelligence—a carnivore must discern what prey is ideal, where to find it, and how best to stalk it while evading detection. Tigers excel at all of the above, thanks to skills acquired during a lengthy tutelage with their mothers. Cubs typically stay with their mothers as long as two and a half years, during which, under her ceaseless care, they learn the many, complicated tricks of the trade. And tricks, at least according to some sources, they most definitely are. During the British Raj, hunters took note of tigers that could imitate the sound of the sambar deer—what naturalist and tiger observer George Schaller would later refer to as "a loud, clear 'pok,'" although he admitted to having seldom heard them make the noise while actually hunting. In colder climes to the north and east, there exist tales of tigers imitating the calls of black bears, ostensibly so they could snap their spines and dine on their fat-rich meat. Tigers frequently adjust their attack strategies to fit their quarry, and whether it's chasing larger animals into deep water where they are easier to kill, snapping the leg tendons of wild buffalo to bring them down to the ground, or flipping porcupines onto their bellies to avoid their sharp quills, tigers are quick studies in the arts of outsmarting their prey. This intelligence, coupled with their innate athleticism and sizable frame, makes for one exceptionally effective natural predator.

Indeed, when one considers the raw physics of a collision with a five-to-six-hundred-pound body moving at forty miles an hour, the equation starts to feel less like one belonging to the natural world, and more akin to that of the automotive. Only *this* Subaru is camouflaged, all-terrain, and has one hell of a Klaxon—not to mention a grill bristling with meat hooks and steak knives. And when it comes to putting a tiger in its tank, this high-performance vehicle runs almost purely on meat—sometimes as much as eighty-eight pounds of it in one sitting. It has its own favorite sort of prey, the hooved, meaty mammals that graze in its domain. But a hungry tiger will eat almost anything.

Of course, there are the more pedestrian items on a famished tiger's menu. Turtles, fish, badgers, squirrels, rabbits, mice, termites—the list is long and inglorious. But then there are the more impressive items that a tiger may take as quarry when the conditions are right. In addition to bears and wolves, tigers have been documented ripping 15-foot crocodiles to pieces, tearing the heads off 20-foot pythons, and dragging 300-pound harbor seals out of the ocean surf to bludgeon on the beach. Bengal tigers in northern India are known to have killed and eaten both rhinos and elephants, and while they tend to prefer juveniles for obvious reasons, full-grown specimens of both species have been victims of tiger predation. In 2013, a rash of tiger attacks upon adult rhinos occurred in the Dudhwa Tiger Reserve in northern India, with a 34-year-old female rhino—almost certainly over 3,000 pounds—being killed and eaten. In 2011, a 20-year-old elephant was killed and partially eaten by a tiger in Jim Corbett National Park, and in 2014, a 28-year-old elephant in Kaziranga National Park farther east was killed and feasted upon by several tigers at once. Keep in mind, a mature Indian elephant can weigh well over five tons; the Bengal tigers responsible essentially took down something the size of a U-Haul truck just so they could gnaw on it. Oh, and lest we forget—tigers eat leopards too, fearsome predators in their own right. Among the most muscular and ferocious of predatory cats, leopards are themselves capable of downing animals five times their size, and hoisting their huge carcasses high up into the trees. However, that doesn't seem to discourage Bengal tigers from crushing their spotted throats and dining on their innards.

But of all the wide variety of flora and fauna the tiger habitually kills, all the Latin dictionary's worth of taxonomy it is willing to regularly gulp down its gullet, there is one species that is notably and thankfully absent: Homo sapiens. Perhaps it's our peculiar bipedalism, our evolutionary penchant for carrying sharp objects, or even our beguiling lack of hair and unusual smell. For whatever reasons, though, *Panthera tigris* does not normally consider us to be edible prey. As we know, they go out of their way to avoid interacting with our kind. But as many a tiger expert has noted, what tigers normally do, and what they're capable of doing, are two very different things. And in the case of the Champawat Man-Eater, normality seems to have vanished the moment our species stole half its fangs—a transgression that the tiger would repay two hundred times over in Nepal alone.

CHAPTER 2

THE MAKING OF A MAN-EATER

Long before an emboldened Champawat Tiger was terrorizing villages and snatching farmers from their fields, it was a wounded animal convalescing deep in the lowland jungles of western Nepal, agitated, aggressive, and wracked with hunger. And it is a safe bet that its first attack occurred there, in the rich flora of the *terai* floodplain, the preferred habitat of the northern Bengal tiger. The *terai* once was—and still is, I discovered, in some isolated areas—a place of enormous biodiversity and commanding beauty. Dense groves of sal are interspersed with silk cotton and peepal, imposing trees that look as old as time. Islands of timber are encircled by lakes of rippling grasses, their stalks twice as high as the height of any man. Chital deer gather at dusk along the rivers, wild pigs root and trundle through the leaves, and even the odd gaur buffalo can make an appearance, guiding its young come twilight toward the marshes to feed. But there are people who make their home here as well: the Tharu, the indigenous inhabitants of the region who lived in the *terai* then as some still do today, in close proximity and harmony with the forest. Residing in small villages composed of mud-walled, grass-thatched structures, and combining low-impact agriculture with hunting and gathering, the Tharu are experts not just at surviving but *thriving* in a wilderness where few others can. The spirits of the animals they live beside are worshipped, and the largest of their trees are as sacred as temples. In short, they are a people with tremendous respect for and knowledge of the natural world. And the Champawat Tiger's first victim was almost certainly one of them.

A woodcutter, possibly, or someone harvesting grass for livestock. A worker whose stooped posture resembled an animal more than a human. Perhaps he was a *hattisare*—a Tharu working in the royal elephant stables, on his way into the land's bosky depths to harvest the long grasses upon which the elephants fed. It is a scene still repeated in the forest reserves of Nepal to this day, and instantly retrievable. We can imagine it: the air spiced by the curried lentil *dal bhat* simmering on the fire, and rich with the tang of fresh elephant dung. Our *hattisare* rides through these aromas atop a lumbering tusker, ducking his turbaned head to clear the low-hanging branches of trees, guiding his tremendous mount with gentle prods of his feet away from the stables toward the dense jungle and grasslands beyond.

The forest is still a wild place here, despite the farmland and pastures being cleared on its fringes, despite the outsiders from the hills who are beginning to buy up the land. But he has committed his puja for the week, making offerings to both the appropriate Hindu gods and Tharu spirits, and besides, he has lived and worked here all of his life—he is a *phanet,* a senior elephant handler with decades of experience. The *terai* has been good to him, he has nothing to fear. He loves the creatures here, and he respects their power, always granting them the wide berth that is their due. Respect, yes, but fear? No, that has never been necessary.

His elephant rumbles beneath him, still content from the *dana* of rice and molasses that composed its last meal. He scratches it behind the ear affectionately, and directs it with a grunted command across a shallow river, toward the plains of high elephant grass beyond. Normally, the harvesting of grass is done with his mahout, but today he let the boy sleep in. He likes being alone with his elephant on mornings such as this, riding up front just behind its head, plodding through the blankets of steam that rise from the marshes and cling to the banyan trees. There is something relaxing, almost hypnotic in the slow, seismic gait of the elephant. From his perch atop its neck, he takes great pleasure in watching the swamp deer graze at the water's edge, or catching a passing glimpse of a rhino calf—or, on rarer occasions still, a fleeing tiger. They, like him, call the forest home, and in that he finds no small sense of kinship.

When they arrive at the spot, he gives the elephant the signal to stop and gathers his sickle. He dismounts, with a little help from the animal's forelimb, and steps gingerly over the swampy ground. At the choicest stalks, he stoops over and begins cutting the tough grass with short, hasping strokes, humming as he works. When he has harvested enough for his first batch, he takes a thin rope from his dhoti cloth and begins to tie up the shock, squatting as he knots the twine.

It's the elephant that senses it first—even though the man cannot see his old friend through the tall grass, he hears his uneasy snort and sudden grumble, deep, resonant, and ominous. He knows that sound well, and all too well what it implies. Perhaps it is best to hurry with his task. The last thing he would want is to stumble upon a fresh kill at the wrong moment, although he does not recall any warning calls of chital deer or flocks of waiting vultures.

But then there is another sound. One with which he is also well acquainted, although he has never heard it so close before. A roar, nearer than he ever thought possible. Close enough to make the grass stalks tremble. Heart seizing, his thoughts clarified by fear, he drops the shock, stands upright—and mounts a lightning-quick debate between the competing instincts of fight and flight. But in the end, there is time to do neither, as the realization comes to him with a stark limpidity that *he* is the fresh kill. He has essentially stumbled upon his own death. With a snarl and a snap and a bold rash of stripes, the tiger is upon him. It has attacked, just as it would a boar or a deer. To the wounded predator, the unknown creature it has caught is so slow and so soft, it barely has to try. It is a revelation of sorts, in whatever shape it is that such things are revealed to the mind of a tiger. A quick bite to the throat, and it's all over. There is no struggle. The nearby elephant trumpets hysterically, but there's nothing to be done—the famished tiger is vanishing back into the tall and rattling grasses, to gorge on its feast, its senses galvanized into a frenzy by this entirely new and imminently available class of prey . . .

|||||||||||||||||||||||||||||||

The Champawat's first taste of human flesh almost certainly began with some such scenario, probably around the year 1899 or 1900, although the details of its initial kills, before it arrived in India, are likely to stay murky. Jim Corbett, one of the few primary sources for the early exploits of the tiger, gives nothing in his account beyond the number of its Nepalese victims. And even in the present day, documenting tiger attacks in the remote frontier of western Nepal is difficult at best—many attacks go unreported, and problem tigers only gain recognition in the press when they've claimed unusually large numbers of victims. Not surprisingly, finding tan-

gible evidence of specific tiger attacks more than a century old in
the region is next to impossible. Unlike the United Provinces, just
across the border in India, there was no colonial government to
publish eyewitness accounts or squirrel away records in faraway
archives.

As for the Tharu, who constituted the bulk of the population in
the lowland *terai* at that time, they possessed a culture that, although
abundant in tradition and nuance, was primarily oral—literacy,
except among a privileged few, was all but unknown. Traditionally, tigers were considered royal property, and only relevant to the
government when it came to sport hunting. A man-eater, unlike
a sport tiger, was greeted with relative indifference, and official
"documentation" would have consisted simply of a tiger skin gifted
to the village shikari who killed it. In the Panjiar Collection—one
of the few historical archives available of communications between
the Nepalese government and Tharu communities—problem tigers are only mentioned twice over the course of fifty royal documents. And in both cases, the responsibility to "protect the lives
of villagers from the threat of tigers" was delegated to the local
authorities, with the warning that "if you cannot settle and protect
this area from these disturbances, you cannot take its produce."
Essentially, to the Nepalese government, man-eating tigers were the
Tharu's problem—not theirs. Rather than send in hunters, they
preferred to let the locals handle it.

There is another reason tiger attacks may have gone unpubli-
cized, though, and that was the cultural stigmas that often attended
them. Predation upon man, conducted by a tiger, was almost cos-
mically aberrant to the Tharu people, whose syncretic belief system
represented a melding of both Hindu and older animistic beliefs.
Such attacks represented the unintended overlap of two separate

spiritual spheres, in the unholiest of fashions. Tigers were regarded as the physical manifestation of the power and grace of the natural world—more specifically, of the forest, upon which the Tharu depended above all else. Under normal conditions, the tigers of the forest were seen as benevolent guardians, even protectors. But if their forest decided to send in a tiger to attack a village, then something in the spiritual health of the community was gravely out of order. A problem with its puja offering, a broken spiritual promise, or some other affront to the gods of the natural world grave enough to summon a distinctly striped form of punishment. Accordingly, it was common belief that the spirit of a tiger victim was doomed, at least in some unfortunate cases, to haunt the earthly realm as a bhut—a malevolent poltergeist of sorts capable of causing bad luck, illness, and even death. And since tiger victims were often totally devoured, the essential and extremely complex Tharu funeral rites of cremation and riverine release became difficult to perform, ensuring further spiritual calamity enacted by the bhut. These destructive spirits could take two forms, that of a *churaini* for women, or a *martuki* for men, and only the help of a shaman, or *gurau,* could keep them at bay. So feared were these specters, it was not uncommon for Tharu widows to pass a torch over the mouth of their departed husbands, to invoke their spirit not to return as a bhut, but instead continue on its path toward becoming a protective pitri, or ancestor spirit. And another complex ritual, involving head shaving, ceremonial rings of *kus* grass, and branches of the peepal tree, would be enacted thirteen days later to ensure that the proper progression had taken place. These funeral rites needed to be performed to create sacred balance in the village, and in the case of many tiger attacks where victims were partially or completely devoured, this was not always possible—

resulting in a spiritual hurdle that put the entire community at risk. With this in mind, one can easily imagine a general reluctance among the families of tiger victims to call attention to the attacks, and risk being blamed for any communal misfortune down the line.

These kinds of stigmas have declined somewhat in the *terai* of Nepal and northern India in recent years, as the presence of bhut has slowly transformed from practiced religion to old-fashioned superstition, and man-eating tigers have faded—although not vanished entirely, as we shall see—from cultural memory. When talking to Tharu *guraus* in present-day Chitwan, I found that the perception of tiger attacks as a form of divine punishment does still exist, although they don't attach any bad luck or ill will to the families of the victims, and they would never deny a funeral service if asked. In fact, they believe that the offended god or spirit will often deposit tiger whiskers on the ground around the village as a form of warning, to give the community the chance to come together and mend its ways before another attack occurs. In the Sundarbans of West Bengal, however, where village men still go into the forest to fish and collect honey, and where tiger predation is still a daily threat, the stigma against tiger victims is very much alive and relevant. Many locals refuse to even speak of tigers or utter their name, as they believe words alone are enough to summon snarls and stripes from the mangrove forests. And when tigers actually materialize and do attack, relatives of the victims are often similarly avoided. "Tiger-widows," as they're unceremoniously known, can be considered unholy or tainted, and at times face abuse from in-laws as well as general ostracism in the community. They are frequently treated as a source of bad luck and forced to live in isolation, where they can wear only white saris and must eschew all forms of decoration, including jewelry or bangles. They are barred from most ceremonies and festivals, and allowed to travel roads only under certain hours. Such shunning may sound cruel—particularly when imposed on a person who has already had a loved one killed by a tiger—but it stems from the very real fears of people who are totally reliant on the forest for their livelihood, and who cannot afford to associate with anyone who may have incurred the forest's clawed wrath. The people of the Sundarbans pray and make offerings to essentially the same forest goddess as the Tharu do in Nepal and northern India—although they call her Bonbibi instead of Ban Dhevi—and they rely on her favor for protection from tigers. When that protection fails, they, just like the Tharu, know that something grievous must have happened to have lost her favor. And this is not something one would want to broadcast in any way.

But problem tigers have not vanished from the Nepalese *terai*. They still exist today. And to re-create what the first, harrowing manhunts of the Champawat must have been like, one need not journey far into the past at all.

|||||||||||||||||||||||||||||

When untangling the skein of information regarding the Champawat, the unavoidable point of entry is the sheer number of its victims.* The tiger is alleged to have claimed 200 victims in Nepal, and then later another 236 victims once it crossed into India. That's 436 human lives taken by a single animal. To put that grisly

---

* For those interested in a more detailed examination of documentary evidence, there is an epilogue at the end of the book which lists the various colonial records, newspaper articles, and physical artifacts that specifically mention the Champawat and provide insight into its attacks.

number into contemporary perspective, the entire roster of the National Basketball Association evens out at around 450 players. So essentially—according to most published accounts—the Champawat very nearly consumed the entire NBA. While comparing its statistics with modern-day professional sports teams' numbers may border on the whimsical, the horror and trauma it would go on to cause for the inhabitants of western Nepal and the Kumaon division of northern India at the turn of the twentieth century was viscerally and painfully real.

But if the number seems wholly beyond the realm of possibility, there are some other man-eaters bounding across the pages of history that clearly demonstrate that large-scale human predation is not beyond the capacity of many apex predators. In France, for example, between the years 1764 and 1767, a wolf—or possibly a wolf–dog hybrid—known as the Beast of Gévaudan reputedly killed some 113 people before Jean Chastel, a local hunter, finally shot it and ended its spree. It is a shocking number, but also one that is fairly well documented, thanks to ecclesiastic funerary records from the Gévaudan region. In 1898, a pair of lions known as the Tsavo Man-Eaters temporarily put a massive British railway project in Kenya on hold when they began pulling workers from their tents at night. Accounts vary as to the total number of victims, with some going as high as 135, although scientific tests conducted by the Chicago Field Museum, which has the taxidermied lions on display, has indicated that they probably didn't actually consume more than thirty-five of their victims. And while its own total tally isn't remarkable in size, the rapidity of the infamous shark that terrorized the Jersey Shore in 1916 has earned its status as the original "Jaws." As to whether it was a great white or a bull shark is still debated—but either way, the deadly fish attacked 5 people and killed 4 in less than 2 weeks. And then of course there is "Gustave," a Nile crocodile from Burundi with a reported length of more than twenty feet, a hide pocked with bullet scars, and an apparent taste for human flesh. In addition to the wildebeest and hippopotamus that comprise its diet, it is said by locals to have eaten as many as three hundred people. These may be some of the more publicized examples, but history abounds with similar predators that have taken humans as prey, in numbers that frequently extend into the dozens, and sometimes even the hundreds. Leopards, brown bears, alligators, even Komodo dragons—they all can and occasionally do attack and eat human beings. It's not common, but it does happen.

That tigers are capable of attacking human beings, under the right circumstances, is beyond dispute. We may not be their preferred, or even usual prey, but that hardly means humans never serve as a source of nutrition. We are made of meat, after all. But is the tally for the Champawat Tiger, a number recorded under less-than-optimal circumstances for fact-checking, and larger than that of any other man-eater on record, actually realistic?

The number of two hundred victims in Nepal—as well as the overall tally of 436 victims—is generally cited in most scholarly works as a credible figure. Perhaps not exact, but reasonably close. This is the number cited later by Jim Corbett, the number evidently certified, tacitly or otherwise, by the colonial British government at that time, and this tally, or similar figures, are repeated by modern-day tiger researchers and tiger hunters alike. Nevertheless, there are some who initially greet the number with a fair and under-standable dose of skepticism, the author of this book included. After all, few things beget exaggeration like fearsome beasts, and the Champawat Tiger's alleged butcher bill does certainly test the limits of credulity. A few pundits have even cast doubt on whether an adult tiger could survive on a diet of humans over such an extended period of time, as the Champawat appears to have done. But even with the rough numbers at hand, the math at least does seem to check out. According to the eminent Indian tiger specialist K. Ullas Karanth, a fully grown tiger needs to kill at least one animal weighing 125 to 135 pounds every week to survive. For normal tigers, this would obviously mean a moderately sized ungulate, like a boar or a deer, every seven days at minimum. Given that the average weight of the humans the Champawat Tiger preyed upon was probably close to that range, then it is fair to say that a fully grown man-eating tiger, so long as it maintained its weekly kill schedule, could readily substitute its ungulate diet with a human one and hunt at the same rate. And if we accept that the Champawat Man-Eater was probably active for the 8 or 9 odd years Jim Corbett's account would later suggest, then that would come out to roughly 52 kills a year over the period—resulting in a hypothetical total of between 416 and 468 human victims, a range that the purported total of 436 human victims falls easily into. It goes without saying that such figures are anything but precise—and it's quite plausible that the Champawat still included livestock and smaller wild ungulates in its diet as well, even while feasting upon humans. But the figures do, at the very least, show that its total victim tally from Nepal and India is not at all beyond the realm of possibility for a tiger that has adopted a primarily human diet, at least from a purely statistical point of view.

Tigers, however, have never been ones to pay much heed to statistics, and in order to lend some legitimate credibility to the Champawat's tally, particularly the more obscure Nepalese portion of it, more tangible evidence than that is needed. Indeed, there are analogous and better-documented situations we can use to show that such prolific man-eating is not quite as implausible as it sounds. Plenty of prolific man-eaters are recorded throughout the recent history of South Asia, although to find the most relevant cases, one need not stray far from the Champawat's original hunting grounds. As recently as 1997, a 250-pound female tiger terrorized villages in the Baitadi District of Nepal, just a short drive north of the Champawat's home turf. By the end of January of that year, the cat had already killed some 35 people; by July, that number had climbed to 50. And by November, it had added another 50 on top of that. In total, in a mere 10 months, this lone tiger was able to kill over 100 people before the government finally dispatched it. Many of its victims, sadly, happened to be juveniles and adolescents, which most likely accounts for its accelerated hunting schedule of an average of 2.5 kills per week. (One can only imagine the all but impossible challenge of trying to promote tiger conservation in a place where two to three children are being devoured by a tiger on a weekly basis.) Were this Baitadi man-eater to have continued its spree uncontested for as long as the Champawat did, haunting the edges of villages and the fringes of the forest, snatching young goatherds and women gathering firewood for the better part of a decade, it is not implausible to think that its total count could have approached a thousand.

And just across the border in India, in 2014, a tiger escaped from Jim Corbett National Park and killed ten people during a six-week rampage. That's an average of 1.67 victims a week, over an extended period of time, in roughly the same geographic region where the Champawat once did prowl. And if there's anything more haunting than the sheer number of victims claimed in such

a short span by this contemporary cat, it's the disarming similar-
ity between its attacks and those of the Champawat more than a
hundred years before. The first victim, a farmer in Uttar Pradesh
named Shiv Kumar Singh, was found mauled in a sugarcane field,
the tiger having almost certainly mistaken him for more conventional prey while he was stooped over cutting cane. The next, a
young woman taking a walk at dusk—her name is not mentioned
in the records—was grabbed by the neck and carried off into the
trees. Not long after that, a laborer named Ram Charan went to
the edge of the woods to relieve himself, only to be snatched by the
tiger and dragged away, screaming for his life. His friends heard his
shouts for help and discovered him lying on the ground with the
flesh stripped from his thighs—he died not long after. And following the first three or four kills, which seemed to be cases of mistaken identity as the bodies were not actually eaten, the tiger finally
figured out that our clawless, weak-limbed species was a fine source
of protein, readily available. From then on, the tiger began eating
its new prey, culminating with its final victim, an older man who
was out collecting firewood in the forest when he was attacked.
The tiger managed to consume part of his legs and most of his
abdomen before a band of appalled shovel-wielding villagers scared
it away. And in an almost eerie instance of déjà vu, this tiger too
was female, it too was injured, and its appetite for human flesh also
provoked a veritable whirlwind of hired hunters, elephant parties,
and distraught locals—which only seemed to provoke it further.

And in both of these modern examples—the man-eater of Baitadi and the man-eater of Corbett National Park—the tigers began
preying on humans for essentially the same reasons: loss of habitat,
loss of prey, and injuries to their teeth or paws. Strong evidence,

clearly, that a compromised tiger with a relatively dense population of vulnerable humans within its territory can and occasionally will feed on them for as long as it is able, and at a terrifying rate.

For modern examples of the actual quotidian challenges that a serial man-eater like the Champawat must have posed to nearby villages, one need not look further than Chitwan National Park—currently Nepal's largest tiger reserve, as well as the home of rare one-horned rhinoceroses, slightly less rare leopards, and a trumpeting bevy of wild Asian elephants. Chitwan, like the vast majority of national parks and tiger reserves in Nepal and India, was once a royal hunting ground, used by the Shah and Rana dynasties over the centuries for its natural supply of tigers and elephants—both of which were considered, to varying degrees, royal property. It received national park status in 1973, when the rulers of Nepal first began diverting their efforts away from killing the once-plentiful tigers toward saving the few that still remained. Its status as hunting reserve aside, however, not a whole lot has changed over the last hundred years or so—at least not within the park itself. True, the local elephant stable, or *hattisar*, shuttles far more foreign tourists atop elephants these days than royal hunting parties, and tigers tend to be shot with telephoto lenses rather than Martini-Henry rifles. But beyond that, much is the same. Tharu settlements still dot the edges of the forest, villagers still graze their cattle in the trees and go into the brush seeking fodder and firewood (although not always legally), and the elephant handlers still perform puja offerings to the forest goddess before venturing into her domain. And, as one would expect in a patch of tiger forest hemmed in on all sides by people and livestock, man-eaters do occasionally appear. The methods of dealing with such tigers are nearly identical to those implemented by the Nepalese authorities of yore, complete with beaters, armed shikaris on elephant back, and even a nineteenth-century method for corralling the cats using long bolts of fabric known as the *vhit*-cloth technique, pioneered by the first Rana rulers—the only major difference being that tranquilizer guns are preferred to actual firearms whenever possible. If a tiger can be captured alive, the Nepalese authorities try to do so, condemning the guilty man-eater to a life sentence at the Kathmandu zoo rather than an execution. But in some cases, bullets do become a necessity, with mandates for termination coming—at least until recently—from the royal family itself.

From a statistical perspective, the research of Nepalese tiger expert Bhim Bahadur Gurung provides what is perhaps the most complete picture of how and why the Champawat began to kill humans more than a century ago. By carefully documenting and researching tiger attacks in Chitwan National Park over the course of several decades, he has essentially created an FBI-worthy profile of how wild, elusive tigers can transform under the right circumstances into serial killers. Between 1979 and 2006, 36 tigers attacked a total of 88 people. The average age of the victims was 36, although the range was wide, from a 70-year-old man killed while collecting wild grasses near the forest, to a 4-year-old girl who was attacked in her own home. Among these victims, more than half were cutting animal fodder of some kind—an activity that involved venturing into forested areas and initiating a stooped posture—and 66 percent were killed while within one kilometer of the forest's edge, indicating that tigers were venturing out of the deep forest and hunting on the marginal zones around human settlements. Attacks increased dramatically from an average of 1.2 persons killed per year between 1979 and 1998, to 7.2 killed per year between 1998 and 2006. This rise was due largely to dramatic growth in the human population in Chitwan, from virtually zero in 1973 when the park was established (the families who had lived there were forced to resettle elsewhere), to the nearly 223,260 people living within the park's new, expanded buffer zone by 1999.* The problem was only exacerbated by grazing restrictions that limited use of communal land, and resulted in more frequent human incursions—often illegal—into forested zones for the collection of grass and leaves to feed livestock. This is all strong evidence of the correlation between the collection of forest resources and tiger attacks, with the majority occurring in the transitional zone where human and tiger habitation overlap, inflicted upon a growing human population actively seeking feed for animals or firewood for their homes.

Even more interesting, however, is what we learn about the tigers. Sixty-one percent of the documented man-eaters occupied severely degraded habitats with low prey densities. Of the 18 problem tigers that researchers were able to examine, 10 had physical impairments like missing teeth or injured paws, with 90 percent of these impaired man-eaters also living in degraded habitats. And of the man-eating tigers that left the forest's edge and ventured into villages—the sort of desperate behavior the Champawat too would eventually exhibit—virtually all came from degraded habitats, and *all* were physically impaired. Unusually aggressive non-hunting behavior was also recorded in some of these tigers, meaning they were

---

* While generally lauded as a landmark event in tiger conservation, the creation of Chitwan National Park involved the forced displacement of dozens of indigenous Tharu families who had called the central forest home—a traumatic event that continues to haunt the Tharu communities that live today on the edge of Chitwan's buffer zone. There has been some progress in terms of giving the Tharu access to the central forest for the traditional gathering of food, fodder, and building materials, although it is highly restricted, and continues to be a source of friction between the Tharu community and park officials.

unwilling to leave a kill even when confronted by humans atop elephants, conduct almost unheard of among normal, wild tigers in healthy habitats. Gurung attributes this aggressiveness to increased competition between tigers for limited territory, and to previous negative encounters with humans, who most likely attempted to chase tigers away from livestock kills so they could salvage the fresh meat for themselves. One of these ultra-aggressive tigers killed five people within a few minutes, and then sat beneath a tree for several hours where a sixth person was hiding, roaring and waiting for them to come down—not the sort of performance one would expect from a famously shy and elusive predator.

But this *was* the kind of behavior exhibited by the Champawat—an animal that was impaired, coping with a changing environment, and that had very fair reasons for being aggressive toward humans. Its pattern of killing almost certainly followed those of Chitwan's most aggressive tigers today, as it became accustomed to hunting humans, first on its own territory in the grassy marshes and sal forests, and then later on ours, among grass-thatched huts and mud-walled houses. It would have progressed over time from chance encounters in the deep forest with woodcutters and foragers, to semi-deliberate confrontations on the forest's edge with grass-cutters and herders, to intentioned kills on the outskirts of villages as farmers worked in their fields or walked into the brush to relieve themselves. As the research shows, the most problematic tigers—those with degraded habitats, physical impairments, and aggressive dispositions—seem to lose their fear of people altogether, and this is precisely what happened in the case of the Champawat. The human settlements that dotted the lowland *terai* ceased to be places of uncertainty and danger, as they were for most tigers, and instead became a veritable smorgasbord. And once that happened, a slaughter of unprecedented proportions commenced.

While statistical analysis of tiger attacks may provide a solid understanding of the underlying causes, data alone does a poor job of communicating their attendant horrors. Attacks by man-eating tigers, though rare, are exceedingly traumatic, in almost every sense of the word. The death of a loved one is always challenging for families and communities, but it becomes far more so when that cherished individual has been mauled or even completely devoured by a striped, fanged, quarter-ton cat. And again, there are contemporary examples of tiger attacks in India and Nepal that provide some idea—albeit a very unpleasant one—of what the aftermath of a wild tiger attack entails.

In the case of lethal maulings—attacks where the tiger succeeds in killing the victim, but either changes its mind or is chased away before it can feed—there is a small but extant body of medical literature on what those wounds involve. When tigers attack a human not out of self-defense, but as potential food, they generally approach the victim much as they would their usual prey of four-legged ungulates. A hunting tiger is stealthy—it approaches its target crouched low to the ground on silent, padded feet, and it waits with twitching tail until the right moment to strike. When that instant arrives, the ambush is lightning fast, and usually conducted from the side or the rear. There is sometimes an accompanying roar coincident with the initial strike—and at 114 decibels, roughly twenty-five times louder than a gas-powered lawn mower, what a roar it is. The tiger will generally use its ample claws to latch on to the prey around the flanks or shoulders, and then seek to kill it with a bite to the neck. On smaller prey, the tiger is more than capable of severing or damaging the spinal cord—its teeth are well designed to wedge between vertebrae and inflict catastrophic damage on the tender nerve tissue beneath, which it usually accomplishes quickly, and from the nape. On larger prey, tigers will knock over the animal first, then strangulate it with a choking bite to the trachea, possibly severing a jugular vein in the process. Humans generally fall into the first category, and when a tiger hunts our kind, it goes straight for the spine, although it will sometimes knock over the victim with a blow from its paws or the momentum of its body.

Such was the case of an attack that occurred in the Nagpur Division of India, and was subsequently described in *Forensic Science International* in 2013; an event that bears a striking resemblance to those attributed to the Champawat. The victim, a thirty-five-year-old woman, was foraging for *tendu* leaves in the forest with her husband and a few companions. The woman was left briefly alone while her husband scaled a tree to pluck leaves right off the branches, when shouts of "tiger, tiger" rang out through the brush. Her husband reached her just a few seconds later, and he was able to scare away the tiger by shouting and hurling stones, but it was too late—she was already dead. When her blood-soaked sari was later removed, and an autopsy performed, the examination revealed "four deep puncture wounds" on the nape of the neck resulting in a "complete laceration of the right jugulocarotid vessel" as well as "compound fractures of the C3 and C6 vertebral bodies due to through and through penetration by the canines of the tiger as a result of enormous bite force used in the killing bite at the canines." The spinal cord at these points was "completely lacerated with multiple foci of hemorrhages." In addition to the severed jugular and broken spine, the victim also suffered multiple deep puncture wounds from the tiger's claws on the arms, shoulders, and torso—some almost two inches wide—as well as a fractured right clavicle and a fracture dislocation of the left sternoclavicular joint from the sheer force of the initial blow. In this case, the death was classified as "accidental," which although true in a legal sense, doesn't capture the purposeful nature of a tiger attack. When one sees the heart-wrenching autopsy photo of the four perfectly spaced, quarter-sized holes on the back of the victim's broken neck, one can't help but feel tremendous pity for the family of the unfortunate woman, and shudder at the expertise with which a tiger does its deadly work. Not malevolently, as man so often does, but naturally, with the grace and ease that 2 million years of predator evolution have bestowed upon it.

As to how the tiger can kill so effectively and quickly, we need only remind ourselves of the considerable toolkit with which the tiger is equipped. As we already know, tigers have four canine teeth that can reach close to four inches, and they have a total of ten claws on their forepaws of comparable length. This means that in the first milliseconds of a full-speed tiger attack, a human body must not only cope with a bone-fracturing impact comparable to that of a charging Spanish fighting bull, but also absorb *fourteen* simultaneous stiletto-deep stab wounds—four of which are usually inflicted on the back of the head or the nape of the neck. And that's just the initial attack. If there's any fight left in the grievously injured victim, it can usually be obliterated almost instantly with a fierce, spine-snapping shake of the head, or further flaying with all those bladed claws. Not surprisingly, survivors of actual tiger attacks are few and far between.

But they do exist. Oftentimes, victims of tiger attacks survive either because the tiger is scared away before it can finish the job, or because it is acting in a defensive manner and not a predatory one—in which case the attack is geared more toward deterrence than nutrition (although tigers have been known to eat victims even when the attack was defensive in nature). Both were likely mitigating factors in the 1974 mauling, though luckily not death, of one tiger researcher in Chitwan, Dr. Kirti Man Tamang. At the time, he was perched some fifteen to eighteen feet up in a tree—a distance considered to be safe from tiger attack—to monitor signals from a radio-collared mother tiger dubbed "Number One" by the team. What he didn't reckon on, however, was just how protective a mother tiger can be. Fellow researchers Fiona and Mel Sunquist, who were working in Chitwan at the time of the attack, describe it in the following passage from *Tiger Moon,* as witnessed from atop a nearby elephant:

> Kirti was moving around in the tree, pointing with the long aluminum antenna. He began to speak; then everyone heard the miaow of a young cub . . . Number One exploded out of the grass with a shattering roar. She made one leap up the tree and in a split second was on top of Kirti. He saw her coming and tried to ward her off with the antenna, but she flung it aside without noticing. She sank her claws into his thighs and buttocks and bit deeply into his leg. The force of her acceleration ripped Kirti off the branch and they both tumbled to the ground fifteen feet below . . . No one could believe what was happening. Kirti's wife Pat repeated "Oh, my God," over and over again, her voice rising in hysteria, but everyone else was dumb with shock. Before anyone could move the tigress charged again, her roars blasting through the silence. The elephants spun on their heels and bolted in blind panic ahead of the enraged tigress. Nothing could stop them. Equipment flew everywhere in a wild confusion of screaming and trumpeting.

People clung to ropes or whatever they could find, trying not to be swept off the elephants in the headlong dash through the bushes.

The research team's elephants may have bolted, but a battle-scarred old tusker was on hand that had participated in royal tiger hunts years ago, before they were banned. It had been trained to be fearless around tigers and had few hesitations about going back into the jungle to recover the fallen researcher before it was too late. Dr. Tamang was found to be in shock but still alive, with a "grapefruit-sized" chunk taken out of his thigh and deep claw marks raking his legs and buttocks. By tiger standards, this was a relatively mild attack—a defensive swat by a mother to deter an over-curious researcher—and yet it still cost the poor man an emergency medical flight to Kathmandu, multiple skin grafts, a nasty bacterial infection, and five full months of painful recovery.

The attack may have involved an outside researcher, but the vast majority of human–tiger conflict occurs among local populations, in the tight-knit rural communities that tend to border tiger territory. And when they do occur, there is a considerable and understandable amount of confusion, heartache, sadness, and anger. A regrettable human tragedy, no matter how you look at it. Hemanta Mishra, a Nepalese biologist with a focus on tiger conservation, was responsible for capturing a number of man-killers in Chitwan National Park, and encountered the sites of recent attacks on multiple occasions. One incident, which occurred in 1979 in the Nepalese village of Madanpur, involved a beloved local schoolteacher who was killed by a predatory bite to the neck. A crowd of villagers was able to scare the tiger away, however, and the schoolteacher's body was saved from being carried off and eaten.

After finally assuring a furious mob that he would deal with the problem—after all, the protected tigers were technically still considered government property in Nepal, just as they had been a century before—Hemanta Mishra describes the following scene:

> The disfigured body of the schoolteacher was lying flat on the ground, facing upward. His mutilated face was covered with dried blood. A group of the dead man's relatives squatted around his body, mourning the unprecedented tragedy. They were surrounded by a large crowd of villagers, silently lamenting the tragic loss of their only schoolteacher. The scene was somber, sorrowful, and silent. The aura of death hovered in the air. From a nearby hut, the wailing of the schoolteacher's wife weeping in pain with her two children periodically broke the silence. A white blanket of cotton and a freshly cut green bamboo bier were laid next to the body. The dead man was a Hindu. His death ritual demanded that he be wrapped in the shroud of white cotton, fastened to the bamboo bier, and transported to the cremation site on the banks of a river. The scene [was] both heart wrenching and gruesome—reminiscent of a nightmarish movie.

Though shaken by what he had witnessed, and uncertain of his ability to actually capture the man-eater, Hemanta Mishra did keep his promise to the people of Madanpur—he eventually shot the responsible tiger with a tranquilizer dart and carried it via elephant to a waiting transport cage, and later, an enclosure at the Kathmandu zoo, where it lived out the rest of its days eating goat legs and chickens instead of human beings.

As disturbing as such attacks can be, the above are not the

worst cases. The results can be far more gruesome when a man-eater is *not* scared away or interrupted before it has begun to feed. The tiger's preferred method of feeding is to drag its fresh kill into a secluded part of the forest, feast on the meat until it can stomach no more, rest for a spell nearby, drink water, and then return to the carcass to continue feeding. It is this behavior that enables trackers to find tigers with bait—once the cat has made the kill, it will generally linger around its prey for several days—but it also means that once a body has been taken into the forest by a man-eater, it is very seldom recovered anywhere near intact. Take, for example, another of Hemanta Mishra's accounts of surveying a kill site following a man-eater attack in Nepal in 1980, involving a cat dubbed "Tiger 118":

> Except for the skull and part of the victim's lower leg, the tigress had eaten almost all of the man. An iron sickle glowed in the bright sun next to the victim's toes. A Nepali topi—a kind of cap—and some bloody rags of clothing were scattered all over the kill site. With a wrenching heart, I watched the two villagers collect the remains of their relative and put them in a jute sac.

Far from being an extreme and unusually disturbing outcome, this scene is fairly typical of a full-scale man-eating event. In a scenario that bears an unsettling resemblance to the aftermath of a suicide-vest bombing, it is often only the human head and extremities that remain, scattered about a welter of blood and shredded clothing where the tiger has been feeding. And in some cases, not even that much is left. In one Amur tiger attack that occurred in the Russian Far East in 1997, virtually all that remained after a young hunter was killed in the forest was a pile of bloody clothing, a pair of empty boots, a watch, and a crucifix. The actual physical remains—a few splinters of bone and bits of flesh—could have fit in a coat pocket. One can only imagine what it is like for friends and family having to contend with the fact that their loved one is not only dead, but actually ingested by an oversized predator still at loose in the forest. And as already mentioned, in western Nepal and northern India, where both Hindu and Tharu funerary rites were closely observed, the lack of an intact body served as a spiritual sort of insult to injury, making the catastrophe that much more traumatic.

Even more traumatic still, however, is the possibility that a man-eater might return—that such a tiger may have acquired a taste for its new prey and actually begin seeking humans out on a re-occurring basis. In these instances, attacks change from chance encounters in the forest to the deliberate stalking of villagers and even predation within their homes. Man-eating leopards are more famous in India and Nepal for dragging victims from their houses, but tigers have been known to do it as well. In addition to the previously mentioned tiger attacks, Hemanta Mishra also relates in his memoirs an attack that occurred in the Madi Valley of Nepal, by a man-eating tigress known as Jogi Pothi. Like the Champawat, this tigress had ceased being an elusive, nocturnal predator and began conducting raids on the edges of villages in broad daylight. And also like the Champawat, this tiger proved extremely difficult to find or catch, as it had a knack for concealing itself immediately after a kill in nearby ravines. The houses of the villagers tended to be simple mud, wood, and thatch structures, economical but not terribly sturdy, which meant that a tiger could break in and drag its victims from their homes. This was very nearly what occurred in the village of Bankatta in 1988. A local yogi—an ostensibly celibate holy man—happened to be furtively entertaining feminine company in the wee hours of the morning when he thought he heard a knock at the door. His "guest" made the mistake of answering said door, as described in the following account:

> Upon hearing the knocking sound, the jogi's lady friend peeked through a hole in the wooden door. Shocked to see a huge tiger, she shrieked "Bagh! Bagh!" ("Tiger! Tiger!") in terror at the top of her lungs. Her jogi consort jumped out of his bed and joined her, banging pots and pans in the hut and yelling for help. Their cries rang across the forest to the village. Equipped with axes and khukuris, Nepalese machetes, villagers rushed toward the jogi's hut, causing the tiger to flee into a nearby ravine.

The yogi's reputation as a holy man may have been ruined, but both his own life and that of his guest were preserved, and the tiger was scared away before it could force its way into the house and complete the kill.

If the thought of a man-eating tiger bursting through wooden doors or mud walls to drag away a sleeping victim isn't sobering enough, there are stories of Bengal tigers braving water and currents to carry off people from their boats. In the aforementioned Sundarbans, a region famous for its unusually aggressive tigers, the cats have been known to swim out and snatch people from their vessels. Despite the mangroves being officially off-limits, locals still do enter into the protected forests to cut firewood and poach animals, activities that put them at risk from a dense population of environmentally isolated tigers with a limited food supply. Inevitably, human–tiger conflict follows. That was precisely what happened in 2014, when a sixty-two-year-old man from the village of Lahiripur set off in a boat with his two children to catch crabs on a small river in the forests of Kholakhali. In this instance, the stalking tiger leapt from the bank of the river, over the water, and into the boat, where it immediately attacked the father. The man's son remembered the tragic attack vividly, as reported by *The Times of India*:

> Suddenly, my sister cried out: 'Dada, bagh (tiger)'. I was stunned, and my body froze. All I saw [was] a flash of yellow. It took me a moment to register the gruesome sight before me. My father was completely buried under the beast. I could only see his legs thrashing about. I shook off my numbness and grabbed a stick. Molina, too, took out a long cutter we use to clear foliage in the jungle. Together, we poked and battered the tiger, but it refused to give up . . . It jumped off and landed on the bank in one giant leap. We saw it disappear into the jungle with my father still in its jaws.

Indeed, tigers do not share the common house cat's fear of water, and at times, they can even incorporate it into an attack strategy. The renowned filmmaker and tiger expert Valmik Thapar took note of how one tiger he observed in India's Ranthambore tiger reserve had mastered the technique of chasing sambar deer into a lake, where, once they were hampered by the water, it would drag them under and kill them beneath the surface. Something similar may have occurred on a human target in the Sundarbans in 1997, when a man named Jamal Mohumad narrowly escaped a watery death. This is his version of the attack, which occurred while he was fishing:

> The tiger lunged at me with its paws. It dug its claws into my legs and dragged me under the water. I struggled under the water and dived down about 10 feet under the water. The tiger let go of me. I swam deep under water as fast as I could. After a while, when I reached the surface of the water, I couldn't see the tiger. I swam down the river for a bit and saw a boat and cried out for help.

Jamal became something of a local legend in the Sundarbans, as he was perhaps the only person on earth who had survived three—yes, *three*—separate predatory attacks by tigers. Despite his harrowing encounters with the animals, he would continue to venture into the forest, driven by the same need for food, firewood, and animal fodder that would have compelled the Tharu people a century before. But in the case of the Champawat, this tiger was no longer content waiting for humans to come passing by. It had begun, by the first few years of the 1900s, to leave the protection of the forest and go out looking for them, undergoing as it did so the transformation from a killer of men, to an eater of men, to an active hunter of them. And in its quest for fresh kills, it would eventually travel away from the marshy grasslands and dense sal jungles of its birth, and begin wandering northward and ever upward, into the populated hills that lay beyond.

CHAPTER 3

A MONARCH IN EXILE

The Nepalese beginnings of the Champawat Tiger's man-eating career may be short on documentary evidence, but it isn't lacking altogether—particularly in a culture so firmly grounded in oral traditions. And ironically, it is in fact a former tiger hunter, not a historian or academic, who appears to have uncovered a convincing report of its early exploits. Peter Byrne (born in 1925) has long been something of a living legend—an admittedly colorful Irish ex-pat and former game manager for the Nepalese royal family, who witnessed firsthand the legendary tiger hunts of yore, before finally turning his energies toward tiger conservation. One of the few Europeans to have gained access to the royal traditions of the *bagh shikar* (the story goes that he won the favor of the Nepalese elite by taking their side in a bar brawl), he was given a rare membership in the postwar years to the confraternity of game wardens and shikaris that served at the pleasure of the Nepalese king. And it was thanks to this intimate familiarity with Nepalese tiger hunters that he first heard stories of "the Rupal Man-Eater," a tiger that devoured scores of villagers in western Nepal at the very beginning of the twentieth century. He was even able to acquire a firsthand account, from the aging father of one of his Nepalese friends—an elderly gentleman named Nara Bahadur Bisht. The ninety-three-year-old man shared with Byrne boyhood recollections of how the tiger had terrified his village, and the massive hunt that was eventually organized in Rupal to stop it.

A glance at the map reveals two eye-opening facts: first, that the Nepalese village of Rupal is just across the border from the Indian town of Champawat. And given the timing, as well as the added details of the armed local response—details that corroborate almost to a tee an account Jim Corbett would later provide—it seems all but undeniable: the Rupal Man-Eater and the Champawat Man-Eater were one in the same. Two names for the same tiger, a Nepalese sobriquet acquired first, and its Indian moniker applied thereafter.

And second? That the village of Rupal is north—*surprisingly* north—of the prime tiger habitat of the deep *terai*. When one looks at the tiger reserves that exist in Nepal today—Chitwan, Bardia, Banke, Shuklaphanta—it is not a coincidence that they seem to cluster, like green beads on a string, along the tropical floodplain at the base of the Himalayas known locally as the *terai*. This is because prior to being deemed national parks, they were royal hunting reserves, kept by the kings for the hunting of tigers. They were chosen specifically as hunting grounds because they were prime tiger habitat, with dense populations of Royal Bengals. This was where the striped cats were to be found, and where they were naturally suited to live. Even in modern times, tigers still cling to the marshes and grasslands of the lowland *terai* rather than venturing into the colder and dryer hills. A 2014 study sponsored by the World Wildlife Fund that measured the tiger population in Nepal found the highest tiger densities "were concentrated in areas of riverine flood plains, grasslands, riparian forests and around wetlands . . ." In Shuklaphanta, tigers much preferred the marshy banks of the Mahakali River. Meanwhile, the dry hardwood forests of the bordering hills supported very low numbers of tigers, primarily because they also offered comparatively low levels of prey. Grazing deer and the tigers that fed on them kept to the rich grasslands and humid jungles of the *terai* floodplain below. It was simply a warmer, greener, and more biodiverse habitat, and it is almost certainly where the Champawat's life began.

Rupal, however, where our tiger would first make a name for itself as a man-eater, is not in the lowland *terai* at all. It's farther north, beyond the first Siwalik hills, in the beginning of the actual *Mahabharat Lekh,* or the Lesser Himalayas. It is a harsh realm of jagged cliffs and bristling pines; a place where the winters are frigid and large animals are scarce. If we assume—and it does seem like a relatively safe assumption—that the Champawat's origins lie in the prime Bengal tiger habitat farther south, in the lush lowland sal jungles of what is today the Shuklaphanta reserve, the obvious question arises: What drove it away from its birthplace, northward into the steep valleys and rugged foothills of the Himalayas, to kill humans on an unprecedented scale? After all, injured tigers with damaged teeth or paws were not unknown in Nepal, nor were man-eaters entirely unheard of. But in the case of the Champawat/Rupal tiger, something without antecedent appears to have occurred. Its presence at that altitude seems almost as unlikely as Hemingway's leopard on the side of Mount Kilimanjaro. What, exactly, was it doing in such an unwelcoming environment?

In answering the question of why a tiger would leave its natural habitat in the *terai,* it only makes sense to look at what was happening in the *terai* at that time. What becomes evident is that the long-standing dynamics between this ecosystem and the human beings who lived within it were undergoing seismic shifts in the late nineteenth century. The deforestation of the *terai* and the displacement of indigenous Tharu people is often attributed to the eradication of malaria in the 1950s via chemical spraying—and there certainly is considerable truth to that attribution. But what history, taken with a healthy dose of analysis, reveals is that while the delicate threads that bound the *terai*, the Tharu, and the tigers may have unraveled almost completely in the twentieth century, they were already frayed long before—as early as the mid-nineteenth century, when the policies of the new Rana dynasty began to take hold. And the early damage done to those intertwined and interdependent cords goes a long way in explaining the emergence of a tiger like the Champawat. When those strands came undone, they released a man-eater like none other upon the world.

||||||||||||||||||||||||||||||

Some 50 million years ago, when the miacid ancestors of all cats were still scurrying through the treetops and the Paleocene Epoch was still in full swing, a tremendous collision took place. The continental plate of India, which had been an isolated island since drifting away from Africa more than 100 million years before, slammed into the Eurasian Plate. "Slammed" in the geologic sense, as it was a slow-motion impact by human standards, occurring at a speed of less than fifteen centimeters per year. But it was dramatic, nonetheless, in the mountain range it eventually produced: the Himalayas. The world's tallest and youngest mountains, born of a buckling that only a head-on collision between continents can provide. This upward thrusting of the earth's crust would eventually engender a bristling range of peaks, reaching well over twenty thousand feet in height, stretching some fifteen hundred miles across, and spanning what is today Pakistan, India, Nepal, Bhutan, and China. From their snowcapped heights, these mountains would in turn beget three major rivers: the Indus, the Ganges, and the Tsangpo-Brahmaputra. The name Himalaya means "abode of snow" in Sanskrit—an ancient Indo-European language that serves as the sacred mother tongue of Hinduism, and that has existed in the subcontinent since at least the second millennium B.C., when its earliest speakers began pouring in from the west. They were hardly the first ones to call the mountains home, however. In what is today Nepal, in the Kathmandu Valley, archaeological evidence has been found that suggests human habitation in the region for at least eleven thousand years. The new Sanskrit-speaking Indo-Aryan arrivals lived right alongside preexisting populations, and in many cases mingled, creating a patchwork of ethnic groups interspersed throughout the range's peaks and valleys.

In some instances, the Indo-Aryan groups who arrived in the Himalaya region—relative newcomers in the grand scheme of things—clung to geographies they were familiar with, while avoiding those that were beyond their ken. They effectively left such domains to the indigenous inhabitants who predated them, while still technically incorporating them into their burgeoning kingdoms. In few places was this practice more pronounced than in Nepal. In the foothills and mountains of the Himalayan range, a series of Hindu kingdoms arose, beginning with the Thakuri dynasty, who ruled parts of Nepal up until the twelfth century; the Malla dynasty, which held dominion until the eighteenth century; and the Shah dynasty, which unified a number of the region's warring kingdoms into a single Gorkha state in the late eighteenth century. These mountain dynasties spoke a host of Indo-Aryan languages, including Nepali, and embraced the tenets and traditions of the Hindu faith, caste systems included.

One thing they did not do—at least very often—was leave the hills for the marshy grasslands and jungles below. This was the *terai*, the rich northern floodplain of the Ganges, a green belt of land that ran a verdant course along the southern base of the Himalayas. The word "*terai*" itself is an Urdu term meaning something akin to "marsh" or "basin," and this is a fairly accurate description for much of the territory. Vast expanses of elephant grass—which can reach up to seven meters—covered wide swathes of the damp ground, and provided ample habitat for deer, rhinos, sloth bears, wild elephants, and of course, tigers. These flat, rippling grasslands were cut through by tributaries of the great rivers that flowed down from the mountains above, and were interspersed by dense patches of forest (aka, jungle), marked primarily by the famous sal trees, which in the *terai* were able to keep their leaves throughout the year. As one might expect, the soil in this floodplain was exceedingly fertile, and with the proper irrigation systems could produce considerable yields of grains like rice and millet. Yet the Hindu Pahari people—who inhabited the hills and mountains above—were generally reluctant to visit the flatter, wetter lands below, for one convincing reason: the entire region was infested with malaria. Whatever agricultural promise it held was offset by the very real risk of contracting a potentially lethal blood parasite. To go into the *terai*, particularly during the warmer monsoon months of the year, was considered a near–death sentence for the people of the Nepalese hills. As the colonial forest surveyor Thomas W. Webber noted in 1902, "*[P]aharis* generally die if they sleep in the Terai before November 1 or after June 1." And even in the cooler months, the risk of contraction still existed. The presence of malarial mosquitoes throughout much of the year provided a natural deterrence against any sort of large-scale settlement. Living year-round in the *terai,* for most Nepali people, was simply out of the question.

There was, however, one group that felt remarkably at home in the *terai*: the Tharu, a people who predated the arrival of the Indo-Aryan Hindus, and who had developed over many centuries a genetic resistance to malaria. They were able to not only survive but thrive in the tropical lowlands, living off the land in small family-based clan units. While their Pahari neighbors clung to their dense villages and terraced fields in the mountains to the north, the Tharu lived in relative isolation in the jungles and grasslands of the *terai* belt below, with small communities strung all along its verdant length. There existed—and continues to exist today— some differences in terms of languages, traditions, and religious beliefs among the various Tharu groups. Many eventually adopted the Indo-European languages of their neighbors, and some, like the Rana Tharu of far western Nepal, even hesitate to label themselves Tharu at all, and insist instead that they are descended from an ancient Rajput king. However, one thing the Tharu all share, from the Rana Tharu of the far west, to the Chitwania Tharu in the central region, to the Kochila Tharu of the east, is a common identity as a "people of the forest." Their own sense of self is intimately and inextricably linked with the natural environment of the *terai*. It is their mother, and it is their home. And for most of the nineteenth century, they depended upon it for virtually every facet of their existence.

This isn't to say, however, that they had no effect upon or interaction with the environment. They most certainly did. There is an increasingly antiquated notion that indigenous peoples engaged in sustenance-based survival strategies exist in a sort of innocent and Edenic bliss within an ecosystem. But with the Tharu, as with people just about anywhere, this was simply not the case. The Tharu did create irrigation canals to yield better harvests from their fields. They did engage in a slash-and-burn system of grass husbandry to feed their animals, not least of which were the elephants they caught and domesticated. And they definitely did cut down trees for timber when needed, and clear space for fields in the forest when advantageous. But they did so with the knowledge firmly in place that the forest could serve as both a natural and a renewable resource. To destroy the forest and the animals that lived within it would have been a form of cultural, if not literal, suicide. They relied upon it for building materials, for firewood, for animal fodder, and for a host of wild foods that they could only find there. This included game such as deer, boar, and rabbit, as well as fish and the freshwater *ghonghi* snails that served then, as they still do today, as the unofficial national dish of the Tharu people (and that taste exceptional, I discovered, thanks again to my host, Sanjaya, when paired with moonshine *rakshi* liquor and served with an eye-wateringly hot ginger-curry sauce). Edible ferns, mushrooms, and wild asparagus were gathered on a daily basis, and a host of medicinal plants were available when needed. The existing ecosystem of the *terai* provided a veritable cornucopia of materials and provisions necessary for survival—without it, they would have had no homes to live in, no fuel to cook with, no animals to raise, and practically nothing to eat. Keeping the forest intact and productive was a priority above all else.

To this end, the Tharu across the full range of the *terai* engaged in a sustainable form of short-fallow-shifting cultivation, growing rice, mustard, and lentils, and rotating crops to allow the soil to recover between plantings. Being seminomadic, most Tharus lived in low-impact mud and grass structures and stayed at a given habitation site for only a few years, ensuring that no single patch of forest would ever be over-farmed or over-hunted. In western Nepal in particular, where the Champawat was born, the Tharu lived communally in family-based longhouses called *Badaghar*—a collective labor strategy that enabled them to pool resources, maximizing their yields while minimizing their environmental impact. All in all, it was a lifestyle that both demanded and ensured a productive forest.

Needless to say, keeping this system of continual usage and renewal running smoothly was something of a balancing act, and for the Tharu, maintaining equilibrium wasn't just about agricultural practices or labor strategies; it had its spiritual dimensions as well. Though nominally Hindu, the majority of Tharu practiced—and continue to practice—a syncretic version of the religion founded upon older animist beliefs. They worshipped and made offerings to the familiar pantheon of Indo-Aryan Hindu gods borrowed from their hill-dwelling neighbors, but also venerated a vast array of forest and animal spirits that predated the arrival of Shiva or Vishnu to the *terai*. For officiating the ceremonies of the former—funerals, in particular—visiting Brahmin priests were largely relied upon, who came down from the hills in the non-malarial months. When it came to the latter, however, the more traditional, tribal elements of the Tharu religion were always conducted under the auspices of the local *gurau*, or shaman (our word "guru" is derived from the same root). The *gurau* was largely seen as the protector of villages, and the intermediary between the Tharu population and the host of bhut spirits—both malevolent and benign—that inhabited the grasslands and forests that surrounded them. Rather than stone temples, the Tharu relied on shrines within the home containing important idols, as well as ceremonies held at specific forest locations called *than,* where the various animal spirits could be worshipped in the open, at the foot of a sacred tree. The local population was generally served by two types of *gurau,* both a *ghar gurau,* who was something akin to a family doctor, and the *patharithiya gurau,* who became involved in larger issues that affected the village as a whole. For example, an illness in the family attributed to unknown spiritual causes might best be handled by the *ghar gurau,* who would attempt to appease the unhappy spirit and convince it to return to the forest. A plague that was affecting an entire village or region, on the other hand, would be the bailiwick of the *patharithiya gurau.* The process of becoming a *gurau* generally took several years of apprenticeship with an established practitioner. Once fully initiated, the new *gurau,* following a ceremonial contract of service between himself and a community or household, was responsible for protecting that community or household from any form of spiritual imbalance. Such ceremonies involved puja offerings of goats, pigeons, and *rakshi* liquor, and could serve as a shield against everything from house fires and crop failures to attacks by wild animals.

Including, as it were, tigers. The *gurau* was responsible for protecting his community from a number of potentially dangerous species, in particular the rhino, the elephant, the sloth bear, and the leopard. Yet it was the tiger to whom the *gurau* held an especially sacred relationship. To be able to live alongside tigers and communicate with them was seen as the mark of an effective *gurau.* When royal Nepalese hunting parties, both Shah and later Rana, came through the *terai* to hunt tigers, they never did so without asking a local *gurau* for assistance, as it was understood that only he had the power to summon the great cats from the forest. And even today, particularly among older Tharu, there is a belief that a truly powerful *gurau* can ride the tiger and use it to travel between villages at incredible speeds. Some will even swear that they've seen this with their own eyes, and they will describe in detail the sight of a wizened old shaman climbing upon the back of a huge, striped tiger and bounding away through the trees. While speaking with the *guraus* of several villages near Chitwan, I heard reference again and again to "Raj Guru," a recently deceased *gurau* they had all known personally and who, despite a penchant for *rakshi* liquor— apparently he was equally famous for his drinking—was still able to summon tigers at will to reach sick villagers in need. To the outsider, such stories sound incredible, but to someone steeped in the cosmology of the Tharu, they are only logical. Being able to live in harmony with the tiger—even gain mastery of the tiger— represents the ultimate form of spiritual ability, because the tiger was and still is regarded as the ultimate expression of the forest's awesome power. A person who can harness the power of the tiger can, in effect, do anything. To the Tharu, the tiger was never a monster to be exterminated, but a force of nature to be harnessed and understood. The truly great man was not he who could kill a tiger, but rather he who could make peace with it, and good use of its fangs and its claws. He who could channel that power, as it were, into something constructive.

And in their own unique way, that's precisely what the Tharu did. The key to maintaining their own sustainable way of life in the *terai* was to keep a certain healthy balance between forces that were ostensibly at odds. The Tharu relied on wild deer and boar as a source of meat, but both would eat their crops if their numbers became too great. Tigers solved the problem nicely, keeping the ungulate population robust but not excessive. However, if the tiger population became too concentrated, then tigers began preying upon livestock—thus, adequate habitat was needed. And because the Tharu also depended on the forests and grasslands for building materials and animal fodder, they were even further inclined to keep ecosystems both productive and intact. This in turn preserved ungulate populations, which further nourished the tiger population . . . and so on. It was a delicate balancing act of sorts, a chain without a beginning or end, that linked together the humans, the flora, and the fauna of the *terai*. But it was an act of balance that the Tharu excelled at. They farmed their fields, they grazed their animals, they hunted and fished in their forests, and they burned and harvested fodder from their grasslands. And they always did so alongside a healthy population of wild tigers.

This does not mean, however, that the Tharu existed in a state of perfect isolation. Their innate resistance to malaria may have allowed them to inhabit an otherwise uninhabitable stretch of wilderness, but they were not totally cut off from the cultures that surrounded them. Indeed, the lands of the *terai* had been incorporated into the various kingdoms and princedoms of the region for thousands of years, serving as a crossroads among the various states that straddled the boundaries of what is today Nepal and northern India. Siddhartha Guatama—better known as the Buddha—is commonly believed to have been born in Lumbini, some 2,500 years ago, in the ancient Shakya Republic of what would eventually become Nepal. People, products, and religious beliefs traveled in and out of *terai* settlements for millennia, and the inhabitants paid homage and taxes to the feudal lords of the region, in times of peace and war, through a rotating caste of Muslim, Hindu, and Buddhist rulers. But while boundaries and allegiances shifted with time, the daily life and culture of the Tharu remained relatively stable. This proved to be true even when the surrounding kingdoms were consolidated following the conquests of Prithvi Narayan Shah, the founder of Nepal's Shah dynasty. Between 1743 and 1768, from his home base in the mountain kingdom of Gorkha, he conquered neighboring kingdoms one by one, eventually combining them into a single, unified state whose borders more or less correspond to modern-day Nepal. From the mid-eighteenth century onward, the Tharu people of the *terai* valleys, although far removed from the capital in Kathmandu, became subjects of the new Nepalese king.

The relationship took time to develop. The Shah rulers of a nascent Nepal initially saw the *terai* wilderness of their new kingdom as a potential source of agricultural expansion, and they actively encouraged the Tharu to increase their taxable farming output through land grants and incentive packages for local communities. But it did not take long for the Shahs to realize there were even more pressing reasons to keep the forests and grasslands of the *terai* uncultivated and intact, *and* that the Tharu people were far more useful as guardians of the forest than as destroyers of it.

The reasons behind this tactical about-face are complex, but high among them was the preservation of a species whose role in the subcontinent cannot be underestimated, particularly in the pre-industrial era: *Elephas maximus,* the Asian elephant. Long before there were all-terrain vehicles, bulldozers, or heavy artillery, there were elephants, and the kings of the Indus Valley had been using them as such since the Bronze Age. In Nepal, there are records going back to at least the sixth century B.C. of state-sponsored elephant management, as evidenced by a report of a Licchavi king named Manadeva who built a bridge across the Gandaki River solely for the transportation of hundreds of war elephants. Works of Sanskrit literature such as the *Arthashastra* are filled with detailed instructions on elephant husbandry, and the Muslim Mughal Empire relied on the exchange of elephants to cement relationships with their neighbors in the Nepalese hills throughout the seventeenth and eighteenth centuries. Elephants were often decreed as royal property, regardless of their provenance, and when one considers the raw potential of their bodies—both constructive and destructive—the reasons for their regal status become abundantly clear. With males reaching weights in excess of 5 tons, with maximum shoulder heights approaching 12 feet, and with trunks that contain more than 40,000 muscles, the Asian elephant possesses awesome strength coupled with tremendous dexterity. For constructive purposes, these traits could easily be harnessed to fell timber, haul stone, erect columns, and steady walls—all of which were necessary for the infrastructure and ceremonial needs of an expanding kingdom. For *destructive* purposes—well, when it came to warfare, there wasn't much a mounted elephant couldn't do. The siege engines of the day, a fully armored elephant with spikes mounted on its tusks and a fortified howdah tower on its back could also function like a Sherman tank. Able to achieve speeds of up to twenty miles per hour, and covered with a hide that could absorb dozens of arrows and musket shots alike, a trained war elephant was more than capable of breaking even the most stubborn of enemy lines, trampling infantry and skewering cavalry horses on its bladed tusks. They provided an elevated vantage point for commanders, and a well-angled shot for mounted archers and snipers. A full complement of military elephants was essential for any aspiring regional power of the day, Nepal included, and their value was certainly not lost on the Shah kings, who had built their vast kingdom, both literally and figuratively, on the backs of well-trained pachyderms.

There was, however, one facet to elephant husbandry that lent a tremendous amount of import to the preservation of the *terai*: the animals were difficult to breed in captivity. The aggressiveness of rival males coupled with exceedingly long gestation periods—up to two years, in many cases—rendered captive breeding extremely impractical, and made capturing wild juvenile elephants an absolute necessity. And in Nepal, the forests of the *terai* was where wild elephants were to be found. When it came to the capture and training of wild elephants, there was one group of people who had centuries of experience: the Tharu. As elephant handlers nonpareil, the Tharus came to be replied upon by the Shah dynasty for keeping the kingdom's supply of elephants well stocked, and always at the ready. This meant the establishment of the government-sponsored *hattisar,* or elephant stable, in many Tharu communities, which came as part of a mutually beneficial arrangement. The local Tharu could use the elephants for their own farming and construction purposes when needed; however, they would be expected to report for duty when the king deemed their services necessary. This included royal hunts, when the Shah king would visit the *terai* to pursue rhino, bear, leopard, and of course, the most regal quarry of all, the Royal Bengal tiger. Hunting tigers demanded that the shikari be mounted atop an elephant, with a local Tharu always serving as the mahout, or elephant driver. The Tharu, along with their royal visitors, even developed a uniquely Nepalese technique for corralling tigers, known as the "ring" hunt. It involved using dozens of elephants to surround a tiger, before finally entrapping it behind a solid wall of trumpeting tuskers. Tharus were well rewarded by the Shah kings of Nepal for their service during royal hunting excursions, with gifts of land, captive elephants, and even a *pagari,* or a "turban of honor" that the recipient could wear with pride for the rest of his life. Such rewards served to solidify the loyalty of the Tharu people, and in doing so, ensured that the *terai* with its herds of wild elephants would be preserved as well.

There was more to it than just elephants, though. As it turned out, the Shahs of Nepal were not the only ones with expansionist ambitions in the region. The British, under the auspices of the East India Company, had been working to establish a foothold in India since the early 1600s. By 1757, that grip was secure, and the Company had begun controlling its most lucrative trade routes with its own private army, effectively making it the de facto ruling power for much of Bengal—much too close for comfort for the Nepalese kings to the north, and vice versa. As the Gorkha kingdom, under the rule of Prithvi Narayan Shah, gobbled up more and more rival territory to form a unified Nepal, a showdown between the two regional powers became all but inevitable. Eventually, it seemed, the British would invade their competitor to the north.

With only one *small* problem—specifically, the same microscopic protozoa that had enabled the Tharu to live more or less unmolested in the *terai* for centuries. The identical malarial parasite that kept most Nepalese hill tribes out of the lowland jungles could just as easily repel invading Britons. And there was the separate but related fact that the land itself was a natural fortress. The wild grasses could grow in excess of twenty feet, making them all but impassable for anyone not mounted on an elephant. The floodplains were a literal quagmire, sucking in men and machinery alike. And the surrounding jungles were purposefully left wild and without roads, which made navigating their tangles next to impossible for any outsider. When it came to protecting Nepal from colonial invaders to the south, the *terai* itself was better than barbed wire. What was home to the Tharu was certain death for Europeans.

The most vivid historical example of the protective barrier that the wild *terai* could provide was the disastrous 1767 campaign of Captain George Kinloch, just prior to the Battle of Kathmandu—also, incidentally, the battle that finally established the Shah dynasty's dominion over a unified Nepal. The British East India Company, which did not care for the Shahs' expansionist ambitions in the least, sent Captain Kinloch into Nepal to break up the siege of the Kathmandu Valley. The Gorkha army, under the command of Prithvi Narayan Shah, had the entire region in a choke hold, cutting off the British from some of their choicest trade routes. The two powers coveted control of the same lands, and Captain Kinloch responded with typical colonial bravado, assuring his commanders that "from Sidley to Nepaul, the road is reckoned extremely good . . . there is no Rivers to be crossed, nor any hills to be passed." His plan was approved, and with some 2,400 men, he set off from India to defeat his Gorkha rivals to the north. And while he did not immediately meet any uncrossable torrents or unscalable mountains, he did come face-to-face with the hard realities of the *terai*. The Bagmati River had totally flooded the land, thanks to seemingly endless monsoon rains, leaving his troops with the excruciating task of dragging cannons through mud that was, according to Kinloch's own journal, "gullet deep." The path before them left little along the lines of food or provisions, and "no trace of any living creature, except wild Elephants, Tigers and Bears which are here in vast numbers." As to actual encounters with tigers, Kinloch's account leaves us guessing, but it seems safe to say that with their roars reverberating through the jungle night, his men did not get much sleep. And the whole time, the bitter deluge never stopped. Kinloch had unknowingly marched his men straight into a death trap, and what followed over the next three hellish weeks included mutinies, bouts of malaria, and thirteen straight days of famine. Even more disastrous, perhaps, than the unforgiving terrain was the defensiveness of its inhabitants. The local "Chaudhary" headman whom Kinloch had commissioned to supply his men with grain essentially abandoned them, and in regards to the rest of the forest-dwelling Tharu, Kinloch would write, "So extremely troublesome were the Jungle people now become, that had a man only fallen a few Yards behind the rest, he was sure to be cut off in a most cruel manner." Delirious and defeated, dodging arrows and spears, poor Captain Kinloch had no choice but to turn tail and run for his life—or more likely be carried, given his feverish state. And while he did miraculously make it out of the wild heart of the *terai,* the *terai* never truly left him. He died the next year, almost certainly of malaria he contracted in its steaming jungle depths.

It almost goes without saying that had Captain Kinloch encountered the easy march into Nepal he initially envisioned—a country stroll filled with tilled fields and helpful farmers—there's a strong chance he and his men could have defeated the Shahs at Kathmandu, and Nepal as we know it would have never been born. But the wilderness of the *terai* and its indigenous inhabitants stopped them, and this was not lost on the Shah dynasty. From that day forward, an informal alliance was formed—one that would only be reinforced during Nepal's second altercation with the British in 1814, when the Tharu and the *terai* once again helped protect Kathmandu from outside invasion. Granted, the second skirmish with the British didn't go as well as the first, and the Shahs lost considerable territory in the west, including Kumaon. But they clearly understood the value of the Tharu people's allegiance, and they made sure that the Tharu were rewarded accordingly.

Evidence of this alliance can be found today in the Panjiar Collection, an extremely rare compilation of royal communiqués between the Shah kings in Kathmandu and the Tharu, primarily in the more populous, eastern part of the *terai* belt, although several records exist from the western *terai* as well, and serve as good indicators of the general dynamic. Many of the documents are *lal mohar,* effectively royal land grants that bestowed rights of management to the local Tharu leaders—usually for rendering a service or showing loyalty—some going as far back as the elephant campaigns in the 1814 war with Britain. Much of the *terai* may have been wilderness, but it was a *managed* wilderness, and the Tharu were the ones best suited for farming it, populating it, and stewarding it in a sustainable way. The Shahs certainly liked grain revenues, but they prized their wild elephant herds, vast tiger-hunting reserves, and impenetrable tracts of malarial borderland just as much. In order to preserve all of the above, the Shahs generally respected the indigenous hierarchy of Tharu villages, and kept intact the authority of the local *chaudhari,* or village chieftain, often granting him many of the same rights as they would a Brahmin noble in the hills. And in a surprising display of open-mindedness for a dynasty of Hindu kings, they even worked within the indigenous belief system of the Tharu to manage village life, as shown by this 1807 decree issued by the court of Girvan Yuddha Bikram Shah to a local Tharu *gurau*:

> To Tetu Gurau, Belaudh praganna, Dhanauji village: We bestow upon you as nankar jagir the uncultivated, forested and

barren lands of Dhanauji village in Belaudh praganna and the revenue of the area except for the king's share. You are given the lands of Gaharwar in Sajot praganna as nankar (tax-free land). Cultivate and make the land populous and protect the people from the threats of elephants, tigers, evil spirits, disease and epidemics. Enjoy all the production of this nankar jagir land. If you cannot settle and protect this area from these disturbances, you cannot take its produce.

The royal decree is telling not only in that it endows a local Tharu spiritual leader with considerable control over *terai* wilderness, but that it also deems the management of tigers and wild elephants as essentially spiritual matters, something only a *gurau* is fit to do. The village in question sat on the edge of Chitwan—prime tiger- and elephant-hunting ground for Nepalese kings. The fact that the king relied on a local shaman to protect its environs from escaped tigers and elephants, rather than government soldiers or paid shikaris, is characteristic of the larger relationship that existed between the Shah dynasty and the Tharu, one in which the latter were not merely subjects, but partners and pioneers in the management of a crucial part of the kingdom. The Panjiar Collection is replete with similar land grants and decrees, with even a few gifted elephants thrown into the mix, for much of the late eighteenth and early nineteenth centuries.

In reconstructing that period, it becomes clear that the Tharu's relationship with the Shah kings, much like their relationship to the forest, was symbiotic and mutually beneficial. The Tharu's natural malarial resistance allowed them to exist and farm in a part of the kingdom where few others could. Their shifting agricultural methods permitted revenues to be collected without compromising the natural protective barrier that the *terai* provided. They were the first line of defense in the event of a British invasion, and they alone could capture and maintain the elephants that were so crucial to the military might of the kingdom. In exchange for these services, the Shahs granted the Tharu considerable autonomy, bestowing the *chaudharis* with the authority to manage their villages through the *lal mohar*. This was not benevolence on the part of the Shah kings *per se*—after all, they were as interested in raising revenues as anyone else—but rather an effective strategy for securing their own borders in the face of a foreign colonial power they did not trust, and who, by 1816, they had already been to war with multiple times. The *terai* served a strategic purpose, and the Tharu were its natural custodians. It only made sense to keep both intact.

It was a partnership that would last as long as there was a need for that impenetrable border with British India—an arrangement that would come to an end, for all practical purposes, in 1846. In that year, an ambitious young upstart from a prominent Nepalese family, Jung Bahadur Rana, took advantage of inner turmoil within the Nepalese government and seized power in a coup. After orchestrating the Kot Massacre—a bloodbath in which he and his brothers trapped and killed some forty members of the court in the royal palace—he expelled King Rajendra Bikram Shah and assumed total control of the government, installing the king's son Surendra Bikram in the throne as a powerless puppet of the new Rana regime. It is notable that when the exiled Shah king was later recaptured, trying to return to Nepal to reclaim the throne, it was in the *terai,* the Nepali region that had always served his family so well. He would spend the rest of his life under house arrest, and the Ranas would go on to rule Nepal for more than a century.

One of the Ranas' first orders of business, upon deposing the

Shahs and solidifying their grip on Nepal, was to improve relations with the British in India. Jung Bahadur Rana was a reformer, a man who embraced Western technology and admired European institutions. He had spent time in India, and he looked to the British colonial system as a model for "modernizing" Nepal and making it a true regional superpower. While the Shahs had garnered their independence through fierce military aggression, the Ranas sought to preserve it with tactful diplomacy. This inevitably meant strong ties with the British to the south. Jung Bahadur Rana would initiate the thawing of Anglo-Nepalese relations with a royal trip to London in 1850, where he would be lavishly received by Queen Victoria, and attend a banquet hosted by the very same British East India Company that his Shah predecessors and their Tharu allies had massacred in the jungles of the *terai*. He would even set aside his traditional Brahmin prohibitions and took on a foreign mistress, in the form of a high-class courtesan named Laura Bell, a single night with whom was alleged to have cost him 250,000 pounds sterling. As to whether he was simply a john, or an enraptured admirer generous with his gifts, there is some debate, but regardless, he seems to have had few compunctions embracing the West, both literally and figuratively.

The Ranas would return the hospitality and friendly overtures of the British by supplying their own "Gurkha" troops to help quell rebellions that cropped up in northern India, and by extending invitations to participate in royal Nepalese tiger hunts. It was a tradition that would carry on well into the twentieth century, culminating in an incredibly opulent hunt organized for Queen Elizabeth II in 1961, an undertaking that involved the construction of a virtual city in the middle of the *terai*. The British royals demurred and let their hosts shoot the tigers on this particular hunt; however, earlier monarchs had no such reservations in doing the shooting themselves. In 1911, King George V and a party of British nobles took part in a traditional Nepalese "ring" hunt in Chitwan at the behest of the Ranas, and over the course of 10 days bagged a total of 4 sloth bears, 18 rhinos, and 39 tigers. Such shoots were commonplace in Nepal in the late nineteenth and early twentieth centuries, with the customary *bagh shikar,* or royal tiger hunt, transforming from a sacred ritual the Shah kings had once used to strengthen local alliances with the Tharu to a means for the Ranas of cementing foreign relationships with colonial powers.

And it worked. With the new Rana dynasty in power, the British in India not only had a military ally in Nepal—they also had an enthusiastic partner in trade. The Ranas actively encouraged exchange with their new friends across the border, seeking to enrich their coffers and emulate the economic policies of their neighbor—basically, to "modernize" and "optimize" their ancient economy. One of the most obvious paths to accomplish this was to increase the economic output of the rich bottomlands of the *terai*—the Bengal tiger's stronghold.

With the British military threat over, the natural barrier of the *terai* was no longer needed, and the Ranas could exploit it however they saw fit. In the decades following the Rana assumption of power, trade with India blossomed, with the western border serving as a primary point of transit, and British Kumaon becoming a mercantilist hub. A range of products were traded, including dyes, textiles, spices, metal, and cattle. But the two major articles of export from Nepal to India are telling indeed: timber and rice. In just five years, between 1872 and 1877, the annual export value to the Indian Northwest Provinces and Oudh went from what was considered a negligible amount to 3,522,280 rupees—roughly twice what Nepal was importing from India from the other side of the border. Of this export total, 548,193 rupees came from timber, and a staggering 1,225,584 rupees came from grain. And both of these primary products, timber and grain, would have come to northern India directly from the western Nepalese *terai*. The transformation of the *terai* from wild frontier to national breadbasket was already well underway by the 1870s. Naturally, this meant an increased clearing of forests and cultivation of grasslands—both of which had a profound effect on the region.

Of course, this transformation also changed the traditional relationship between the indigenous Tharu and the central government, and hardly in a positive way—it effectively robbed the people of their autonomy and of their stewardship of the land. In the transition of power from Shahs to Ranas, the *terai* transformed from a frontier to be preserved to a resource to be exploited. The old allegiances between the Tharus and the Shah kings—allegiances forged through decades of collaborative elephant hunting, pioneer farming, and resisting the British—simply did not fit into the Rana vision of a modern Nepal. Accordingly, the relative autonomy and self-management over the lands of the *terai* that the Tharu had enjoyed became an obstacle to Rana hegemony. And as such, the Tharu themselves went from useful allies in managing the *terai* to a hurdle standing in the way of political and agricultural reform.

Steps to remove this obstruction appeared as early as 1854, when the freshly empowered Ranas decided to institute the *Muluki Ain,* a legal document that imposed a state-decreed caste system upon the Nepalese, and relegated the Tharu to the inferior, impure status of *masine matwali,* or "enslaved alcohol drinkers"—a legal designation that would persist for more than a century. This affected their rights as citizens, and gave the upper hand to any higher-caste hill dwellers who might happen to deal with them in matters of land or labor. And just seven years after that, in 1861, the Ranas imposed a system of land management called the *jimidari*. The new local headman, or *jimidar*, could *potentially* be a *chaudhari* from one of the more elite Tharu families, although the title was increasingly being granted to the high-caste Brahmins from the hills, especially friends and relatives of the Ranas. Essentially, it was cronyism in its purest form. As Tharus often did not possess official written title to the lands they used, such tracts of forested or sparsely culti-vated land were frequently given away to the entrepreneurial *jimidar* landlord, under the condition that he convert them to rice paddies and recruit tenants to farm them, thus increasing revenues. "Re-cruit" may not be precisely the right word, as the existing *kamaiya* labor arrangements of the time amounted to what most would call indentured servitude, in some cases even slavery. Land grants to *jimidar* were usually *birta,* or hereditary, which fixed this new social structure firmly into place. The old system had broken, and the old allegiances were gone—under the 104-year rule of the Ranas that followed, only three *lal mohar* land grants would be given to Tharus in the *terai,* compared with the dozens they had received during the reign of the Shahs. The land was being handed over to the hill-dwelling elite in exchange for political favors, and they in turn could develop it however they chose.

The effects of this transformation would ripple out, destabiliz-ing the culture, the land, and its animals alike—not as profoundly as the truly ruinous changes that would occur with the chemical eradication of malaria in the 1950s, but damaging nonetheless. A formerly seminomadic people engaged in hunting and gather-ing, while also practicing only a minimally invasive form of forest agriculture, were for the first time becoming tied to the land as indentured servants—essentially, a rural peasantry aimed not at small-scale subsistence farming, but at producing an exportable and commodified surplus. The diverse forest resources that had traditionally sustained the Tharu became increasingly unnecessary to the Rana, and the wild populations of deer, elephants, and tigers they had helped to keep in balance were pushed aside to make way for farmland. The eventual physical effect of this cultural shift was a breaking up of the formerly continuous ecosystem of the *terai,* and replacing it with restricted pockets of residual wild habitat. It is true that the Ranas maintained traditional hunting reserves for tigers, including those in Chitwan, Bardia, and Shuklaphanta, and that they engaged in tiger hunting on a massive scale, in elaborate hunts designed to demonstrate their authority, and later, as we well know, to curry favor with foreign guests. And the Tharu would indeed continue to work at the *hattisar* in such places, patiently awaiting the next royal visit. However, despite their passion for tiger hunting, the Rana rulers seemed more than happy to watch the lands that surrounded the reserves become productive livestock farms and rice paddies. To a dynasty bent on catching up with their colonial British neighbors through economic optimization and agricultural reform, what other purpose could such lands have possibly served?

Certainly not as prime tiger habitat. Beyond the boundaries of the hunting reserves, tigers were nothing but a threat to the Nepalese government's new spirit of economic progress and productivity—a relic of the old Nepal rather than a harbinger of the new. They killed cattle, they frightened loggers and migrant laborers—essentially, they served no useful purpose whatsoever once outside the verdant limits of a royal park. And whereas the Shahs of yore might have relied on a local Tharu shaman to keep his village spiritually balanced and safe from tigers, the Ranas would have surely taken their cue from the British across the border and relied instead on hefty bounties and high-powered rifles. The lives of tigers became imperiled once they ventured into the cultivated lands that lay beyond the reserves—their safest bet was to steer clear of the farmland and villages altogether, and stay hidden in the strips and patches of wild forest that remained.

Which posed one rather significant problem: wild tigers are unbelievably territorial. Males in particular, although females are hardly generous when it comes to the sharing of their space. Unlike African lions, which figured out at some point in their evolutionary past that working together in prides made chasing down prey on the open savanna much easier, jungle-dwelling tigers have always gone it alone, relying more on stealth to find food than on social relationships. With the exception of cub-rearing and mating, they are solitary creatures. The size of an individual tiger's territory can vary widely, depending on the availability of prey and the sex of the cat. But in nearly all cases, the space required has the potential to push the boundaries of what a tiger reserve or an isolated patch of forest can readily provide. George Schaller, over the course of his Bengal tiger research in Kanha National Park in India in the 1960s, estimated the range of one female he studied to be roughly 25 square miles, while that of a male was put at 30 square miles. Fiona Sunquist, during her research at Chitwan National Park in Nepal a decade later, put the average range of females at 6 to 8 square miles, while the ranges of males could be considerably larger, in excess of 50 square miles. And when tigers resort to man-eating, ranges can expand even farther in their quest to find available human prey. Although tigers can and do settle in a particular area if the hunting proves especially good, finding that area can take them on long and winding treks. The legendary tiger hunter Kenneth Anderson recorded man-eaters in the 1950s with ranges that covered the gamut from 100 square miles to 600 square miles, and Jim Corbett himself described one man-eater as having a range of 1,500 square miles. That may sound improbable, at least until one considers that in the Russian Far East, where natural prey is generally far scarcer than India or Nepal, Amur tigers regularly have ranges that extend into the hundreds of square miles, some so large that the cats seem more like perpetual wanderers than settled predators, forever prowling the frozen night beneath burning stars.

Exact numbers aside, one thing is clear: tigers need room to maneuver, and they're seldom willing to share that space once they find it. Which is why even a slightly fragmented natural habitat can be highly problematic. There is, after all, a very good reason wild tigers have an average life span of about twelve years, half the age they often reach in captivity. There is so much competition for territory, older tigers frequently get mauled by younger rivals seeking to establish themselves in an area. The clashes are especially fierce among territorial males, who will stand on their hind legs and commence shredding each other with their foreclaws and fangs, until one of the combatants either dies or flees. But violent encounters occur between females as well. In either case, when a tiger is no longer physically able to defend its territory, it has no choice but to leave. And if available territory is limited to begin with, that usually means that the exiled tiger, cut off from its natural habitat and normal prey, will end up somewhere it's not welcome, scrounging for food in places it does not belong. These scenarios often occur with older tigers, but they also manifest themselves with younger, disabled tigers that have been injured in fights with rivals, or—most relevant for our purposes—wounded at the hands of man.

A prime example of what most likely happened to the Champawat is the case once again of a more recent man-eater in Chitwan National Park. The tiger in question, nicknamed Bange Bhale by researchers, had been a dominant tiger in the park between 1982 and 1984. That was, until a younger, more powerful tiger called Lucky Bhale wandered into its territory and challenged it. Bange and Lucky dueled, and Bange lost big. Injured, Bange limped out of its territory, unable to catch its normal prey thereafter. Not surprisingly, the marginalized cat, left without a home range or an effective means to kill deer, adapted to its new situation and instead became a killer of humans. Bange's first victim was a man cutting grass in the southwest of the park. Its second kill was a man collecting firewood on the western bank of the Narayani River. And its third victim was a fisherman sleeping in his mud-and-thatch house by the water—the tiger broke in and dragged him out of his bed. Its fourth victim, however, was luckier than the first three. An elephant driver, he had a sickle with him for cutting elephant grass, and using it, was able to fend off the tiger long enough for his friend to scare it away. He hardly escaped unscathed, however; during the short tussle, the tiger had lodged its teeth in his face, leaving the man with severe damage to nerves and muscle tissue. The tiger was successfully captured after that and transported, like other man-eaters in the park, to spend the remainder of its life behind bars at the Kathmandu zoo.

It's not difficult to imagine a similar scuffle driving the Champawat Tiger out of its native range in the low *terai*. It could have been a roaring and snarling encounter with a rival female, or, if the Champawat had cubs, even a fight with an adult male, as grown males will try to kill cubs fathered by rivals. Mother tigers are notoriously protective, and they will battle to the death to protect their young. In 1981, in India's Ranthambore tiger reserve, Valmik Thapar recorded a highly unusual encounter between a mother tiger and an aggressive male. Based on tracks found the next day, Valmik was able to re-create the event, which unfolded as follows: The female tiger attempted to distract the approaching male with affectionate advances while its two young cubs scampered away to hide. The cubs, however, made the mistake of returning too soon, only to discover a furious male annoyed at being disturbed and bent on tearing them to pieces. The mother tiger, with a resounding roar that could be heard at a guard post a full two kilometers away, objected. As Valmik would later write:

> It appears that the male must have moved, in a flash, towards the cubs, and the mother was forced to take lightning action. With a leap and a bound she attacked the male from the rear, clawing his right foreleg before sinking her canines in and killing him. It was an amazing example of instinctive reaction: a tigress killing a prime male tiger to save her cubs from possible death.

But the tigress's revenge didn't stop there. The furious mother actually proceeded to "open his rump" and completely devour one of his legs. Convincing evidence, as if any was needed, that tigers do not react well to violations of their space or their property.

Unlike the tigress described above, however, that dispatched the rival tiger with a single bite to the neck, the Champawat did not have a complete set of canine teeth—a serious handicap that would have left it at the mercy of more fully abled cats, not to mention other species. Besides contending with territorial rivals, wild tigers must also occasionally spar with other dangerous animals as well. This can include prey, boars in particular, which can be especially problematic for even a healthy tiger with a complete set of canines. Take for example this account recorded by the Indian Forest Service official J. E. Carrington Turner, which tells a gruesome version of one such encounter:

> The tiger walked down the bank on to the soft sand, and circled the boar who wheeled to face him. The circling and wheeling continued three or four times until suddenly the tiger charged the boar, striking him with all his might, and, with the greatest agility leaping aside after his blow. The boar met the onslaught by turning dexterously and taking it on the side of the shoulder. In this manner the tiger delivered blow after blow . . . his blows told their tale, for the boar was dripping with blood from his shoulders downwards . . . The tiger, in striking and endeavoring to jump away from the boar, either lost his balance and landed awkwardly, or skidded in the soft sand. Seeing his opportunity, and astonishingly quick to grasp it, the boar charged straight into him, and, burying his razor-edged [tusks] in the tiger's belly, ripped and ripped again, reinforcing his thrusts to the fullest by his stupendous might and enormous weight. With the greatest difficulty the tiger succeeded in extricating himself and stood apart . . . He was disemboweled, with much of his entrails dangling low and heavily from his belly. Slinking towards the bank, with dragging intestines, he ascended it and disappeared in the thorny scrub bordering it . . .

And as if potentially lethal prey weren't enough, other species of predator are also quick to take advantage of a compromised tiger unable to properly defend itself. Leopards, wolves, sloth bears, and even wild dogs can all pose potential threats to a wounded tiger, and sometimes even a healthy one as well. Kenneth Anderson, the aforementioned tiger hunter, once observed an entire pack of dogs—nearly thirty total—chase down a tigress and eventually gang up to kill her, although not before she managed to snap the spine of one "with the sharp report of a twig," and claw the life out of five more.

All of which would have proved a serious problem for the Champawat. True, it was still an unbelievably dangerous creature, with the capacity to kill human beings in a matter of seconds. But downing a Homo sapien was an entirely different story from taking on some of the largest and most aggressive animals in the natural world. Unable to defend its territory against rivals and aggressive males, incapable of taking down much of its usual prey or fending off competitive predators, its existence was gravely threatened. It would have been forced out of its home range, to fare as best it could in the human hinterlands beyond, hunting its new bipedal prey. With only one issue: thanks to malaria, the *terai* was still not terribly populated. While logging and agriculture undertaken by absentee landlords had begun to transform the landscape in a decidedly human direction, the Tharu settlements to be found there were still relatively sporadic and widely distributed. In short, for a cat that ate almost exclusively humans, these were not prime hunting grounds at all.

Farther north, however, where the swampy, flat jungles turned to tall, craggy hills, malaria was far less of a threat and human beings settled in denser concentrations. The peaks and valleys were dotted with the villages and terraced fields of the Pahari, or hill dwellers, with plenty of brush in the nooks and crannies to hide in. No tiger in its right mind would pick the outskirts of the noisy towns and nearly deerless forests of the Nepalese middle hills as home, but it had virtually nowhere else to go. To exist in the world meant to leave the familiar elephant grass and sal trees of the *terai* behind forever. Survival, in effect, lay elsewhere, in the misty peaks above. More than a century later, we can still imagine its path, prowling, lurking, ever hunting, heading north through the grasslands, past the ancient stonework of Baitada's temples, along the fringes of Daiji . . . then up the first steep ascent of the Siwalik Hills, golden tiger eyes flashing, on the lookout for fresh meat. After cresting the first ridge, it's down into the inner Madhesh Valley, which offers more in the way of cover, but still little along the lines of anything human to eat. So the tiger continues, its ears cocked to pick up our peculiar sounds, its nostrils flared for our familiar smells. Soon, it's going up again, skirting stands of heavy oak and shaggy pine, into the peaks of the Mahabharat Range, the very beginning of the Himalayas . . .

And then the Champawat finds it. Its own Nirvana. Tucked into the folds of the mountains, in the district of Dadeldhura, is the village of Rupal. A place so perfectly situated for the hunting of human beings, the tiger has no reason—at least no *natural* reason—to ever leave. And like an angel of death, it descends upon the valley.

<hr>

Even today, upon examining Rupal from a bird's-eye view, an obvious truth emerges: the village is, almost literally, a human target. When seen from a satellite's lofty perch, it appears as a populous, beige-tone bull's-eye of houses and cultivated fields, surrounded on all sides by rings of steep, green hills and densely wooded ravines. From accounts of more recent Nepalese tiger attacks recorded

outside Chitwan, it has been well established that the most "successful" man-eaters—meaning those that are able to evade capture
and continue killing on a regular basis—are those that have adopted
the technique of snatching humans from the edge of villages and
quickly dragging them away into steep, wooded ravines, places
where hunters often cannot give chase even if they wanted to. In
this way, a man-eating tiger could finish its kill uninterrupted, and
steer clear of anyone who might attempt to stop it.

In Rupal, it seems this was what our tiger did. With this dense
pocket of human settlement, surrounded on all sides by wooded
gorges, the tiger could essentially attack from any angle, keeping
the populace constantly guessing and in fear. Hemmed in on all
sides, there was nowhere to go—they still had to tend to their
fields, gather wood, and cut fodder on the village edge, all of which
would have put them at risk from a tiger that had mastered the
technique of swooping in from the tree line, snatching them by the
neck, and spiriting them away into the dark depths of a ravine. To a
man-eating tiger, picking off humans from such an ideally situated
settlement would have been akin to shooting fish in a barrel.

Which explains, finally, the puzzling course of the Champawat.
To a normal tiger, the steep, rocky terrain of the Nepalese middle
hills would not have been ideal habitat at all. The lowland *terai*
would have been a much better fit, full of game and bountiful in
potential mates. But to an abnormal tiger with limited prospects
for establishing its own territory, and an inability to catch much
of its natural prey, the *terai* would have no longer sufficed. For a
compromised man-eater, the humans of the sparse settlements of
Tharu would not have been enough to sustain it. Further, with
increased competition for diminishing habitat, it would not have
been able to hold its own against competitive tigers. The hills to

the north, on the other hand, would have provided exactly what a tiger like the Champawat was searching for: unclaimed territory full of human food. It was not coincidence or happenstance that this tiger ended up far from home, prowling the outskirts of Rupal in the towering hills of the Himalayas. Tigers are, after all, intelligent and versatile predators that can adjust and adapt their hunting strategies to suit their environment and prey. The Champawat chose this location because of just how conducive it was to its new mode of hunting. It was, in effect, a strategic choice. And as a strategy, it proved extremely effective.

Too effective, as a matter of fact. For while the tiger was able to evade bounty hunters and paid shikaris for several years, its weekly toll of human victims eventually mounted to the point where something drastic *had* to be done. As to who made that decision, there is some divergence of opinion depending upon the account. According to Jim Corbett, the Champawat was hunted by "a body of armed Nepalese, after she had killed two hundred human beings." Many historians since have echoed that sentiment, inferring that it was a regiment of the Nepalese Army, or a cadre of official government shikaris sent in to finish the job. Weapons were not commonplace among villagers, meaning outsiders would have likely been required to organize and participate in the ensuing hunt. If this was the case—and particularly if elephants were involved on a large scale, as they almost always were in tiger hunts—there must have been at least some government involvement, possibly even a direct mandate coming from the Rana prime minister in power at that time, Chandra Shumsher Jang Bahadur Rana. Rupal was, after all, just a two-day march from one of the Ranas' favorite *terai* hunting grounds, and an escaped tiger that was killing villagers by the hundreds might have been a topic of interest. The occasional man-eater was by no means unheard of, but a single tiger devouring two hundred people was another thing entirely—an unprecedented catastrophe that would have demanded an unprecedented response. A show of force, as it were.

This seems plausible, although the account of Nara Bahadur Bisht, the elderly Nepalese gentleman interviewed by Peter Byrne, and who was a boy in Rupal at the time of the attacks, paints a slightly different picture of the event. According to his recollections, the Ranas turned a deaf ear on the pleas of villagers, ignoring whatever petitions were sent their way. As a result, according to Bisht, the large-scale hunt designed to finish the Champawat was largely a grassroots endeavor; the local villagers, realizing no help was on the way, organized the party themselves, drafting a thousand men from every village for twenty miles, to be summoned upon the next human kill. Such a scenario is certainly possible, although Bisht very well could have been influenced by anti-Rana sentiments that persisted in Nepal in the latter half of the twentieth century, following the revolution of 1951—an uprising that ended the Rana dynasty. In the wake of the revolution, the desire to attribute agency to the general populace, rather than an unpopular leader, is more than understandable.

The truth likely lies in the middle. All accounts seem to agree that the ensuing hunt was extremely large and logistically complex. It is logical that the common beaters, walking at arm's length and making as much noise as possible, would have been volunteers from the local population. And it could have been they, under the leadership of the village headman, who demanded the hunt and organized its logistics. On the other hand, any elephants, armed shikaris, or soldiers involved would have likely been at the very least approved by a government official, if not the Ranas themselves.

Given that the *hattisar* stables were generally semiautonomous between royal visits, it's not improbable to think that elephants trained for hunting would have been drafted, at the very least, with a local official's tacit approval.

As to how the events unfolded, there is no recorded tiger hunt in recent times to compare it with—tiger hunting has been banned in Nepal since 1972, and the royals seldom allowed videos or photographs to be taken in the decades prior. From accounts of older Nepalese hunts, however, and from more recent tiger captures using tranquilizer guns, a picture emerges of just how impressive such an endeavor would have been.

In the first wave, pushing headlong into the trees toward the latest kill, a thousand beaters—many of whom had no doubt lost loved ones to the tiger—all shouting at the top of their lungs and clattering their curved kukri machetes to flush out the beast. Behind them, a bristling phalanx of elephants with armed shikaris on top, battle-scarred tuskers more than capable of stomping a tiger to death, trumpeting and rumbling in their eagerness to do so. And finally, in the rear, an entire company of Nepalese soldiers, rifle-toting Gurkhas, war-hardened and ready to kill should the tiger break free of the first line. A veritable army, assembled and marching, for the purpose of ending the Champawat's reign. Using goats or young water buffalo strung along the valley as bait, they would have been able to close in on the tiger's location. And given the topographical positioning of Rupal, it's not hard to guess at their subsequent strategy. The village is surrounded on all sides by steep hills and ravines, a landscape that is not conducive to the traditional Nepalese "ring" hunt method practiced in the flatlands of the *terai*. There is, however, a single outlet from the valley to the west, leading directly to the Sharda River. With this army of men and elephants forming a U and slowly closing in on the tiger like a massive pincer, it would have been feasible at last to funnel the cat into this narrow pass, and then to corner it against the steep banks of the river. Which, apparently, was exactly what happened.

In trying to re-create the tiger's final moments in Nepal, we can imagine, perhaps even with some semblance of pity, this wild animal that has already had both its body and its instincts mangled by the bullets of man, all at once cornered and confused on the cliffs above the surging river, its eyes darting wildly, its heart hammering beneath its panting flanks. Its sensitive ears, designed for the faint rustlings of hooves in the forest, or fur through long grass, are instead overwhelmed by the collective thunder of a thousand men screaming, a hundred rifles firing, and a dozen elephants blasting their trumpets into the air. It moves an inch closer to the rocks' edge, its claws scraping and clattering for purchase, unable to go a step farther without suffering a great fall. But when it sees them break through the trees, when it spies the puffs of smoke from atop the elephants and the harsh spouts of gravel at its feet, it knows, in whatever way a tiger can, that its only hope lies across the river, in the strange land beyond. And with that it roars and it leaps, a tiger falling, orange and black stripes plunging against the gray bands of rock. It vanishes with a splash just as the first team of shikaris reach the ledge; they fire shots into the rushing water far below, and they continue firing when the tiger finally emerges, soaked and scrambling, on the opposite bank of the river. But it's too late—within seconds it is gone. It is in India now. It is in Kumaon.

Where an entirely new hunting ground—and a new hunter—are both waiting for it.

||||||||||||||||||||

CHAPTER 4

THE FINEST OF HER FAUNA

The first time Edward James Corbett heard mention of the tiger that would come to be known as the Champawat was reportedly in the year 1903. He was then neither a famed tiger wallah nor a tracker of man-eaters, nor even yet a member of the Order of the Indian Empire. He was nothing more than a low-level railroad employee in his late twenties, the son of an Irish postmaster who had never studied beyond high school. The topic came up while on a hunting trip in the forests of Malani with his friend Eddie Knowles, a man who by Corbett's own account "was one of those few, very fortunate, individuals who possess the best of everything in life." Knowles, unlike Corbett, came from a privileged family, holding lofty positions in the Indian Army and colonial society alike. Back in Britain, such an outing that Corbett was engaged in, in Malani, would have been unthinkable—just like Hindus, the English had their own ancient caste system, and "shooting" was a ritual practiced solely by the aristocracy. A blue blood like Knowles would not have been caught dead pacing the family estate in high tweed with a "country bottled" railroad man from the sticks like Corbett. But India was not England, and the vicissitudes of colonial life rendered social boundaries somewhat more fluid.

Jim, as his friends liked to call him, was an exceedingly likable fellow from a hardworking family, and despite his humble background and limited prospects, known throughout the district and respected by all—Englishmen and Indians alike. And more than that, he also was something of a curiosity. As a domiciled colonist born and raised in the hills of Kumaon, he was one of those rare figures who seemed equally at home in two separate worlds. He was as comfortable chatting in Kumaoni as he was speaking "the Queen's," and as at ease tracking sambar through the jungle as he was playing bridge at high tea. The Victorian literature of the day was obsessed with the idea of the "wild child" and the "noble savage," particularly ones with European lineages like Tarzan and Hawkeye. In Jim Corbett, a man who had spent most of his life scouting and hunting through the local forests with indigenous shikaris, well-heeled Britons like Eddie Knowles no doubt imagined they saw some of those romantic notions played out in real life. Corbett could imitate the grunts of a leopard or the chuffing of a tiger with an accuracy that sent a collective shiver through a dinner party; his ability to pick up a spoor in a waterlogged forest put the keenest bloodhounds to shame. In short, in the rugged frontier of northern India, even for a colonial elite like Knowles, Jim Corbett was a useful and interesting friend to have. And besides, when it came to hunting, Corbett was known far and wide as one of the best shots around.

As for the first mention of the tiger, it's not difficult to imagine: the two young men dressed in their field khakis, the first inklings of sweat stains beginning to show at the creases, fowl guns carried lackadaisically over their shoulders. Corbett is lean and wiry, a tad bit shy, Knowles a little less so, on both counts. They share the speech and mannerisms of faraway England, but there is something subtly different about Corbett—a knowing spark behind his expressions, a Celtic twinkle at the corners of his eyes. And despite his prematurely thinning hair and slender frame, there is a sort of toughness as well, a hidden resiliency. It is a body that has known its share of hardship, one that may bend but does not break—and also, as Knowles surely knows from having hunted with him before, a body set to spring into action like a trap. On this day, though, given its overall pleasantness, the two friends are content to take things a little more casually; the afternoon sunlight has turned their search for game into a relaxing stroll. As they walk, they share hunting stories, or "*shikar* yarns," as Corbett likes to call them, and as per usual, Knowles has the best ones. Stories of drunken pig-sticking with British officers, perhaps, or trick shots from atop elephants with maharajas—all of which Corbett listens to with polite but detached interest. For his friend, hunting is purely sport, a source of recreational adrenaline and trophies for the parlor. For Corbett, however, it was how he had helped feed his family following the untimely death of their father. And as one of fifteen children, putting meat on the table had been a considerable task indeed—much of his youth was spent with his brothers scouting for food in the forests of Kaladhungi and Nainital. Still, Corbett listens to his friend and nods along to the tales, even when they become predictable and his mind begins to wander.

But then Knowles mentions tigers. One tiger, in particular, and Corbett's ears perk up like a swamp deer's in alarm. *Two hundred?* The number sounds incredible, and he asks Knowles to repeat it. Knowles confirms the figure and sketches the outline of the story so far. Two hundred people in Nepal, give or take a few. And even a company of armed Gurkhas, in all their legendary fearlessness, had failed to stop it; all they had managed to do was chase it across the river, into Kumaon. Naturally, Corbett has heard tales of man-eating tigers—who in the British Raj hadn't? But never had there been one so prolific or so close. The tiger had been wreaking havoc along the entire eastern border. They'd set bounties, hired shikaris—even dispatched troops from Almora.

Knowles, however, is confident that the creature's days are numbered. He tells Corbett that the government has sent in the best man for the job—his own brother-in-law, B. A. Rebsch from the forest department, the finest tiger shikari in the world. In fact, they deputed him just for the occasion.

|||||||||||||||||||||||||||||||||

Jim Corbett grew up within earshot of tigers' roars, and as an expert hunter who had spent his formative years tracking alongside indigenous shikaris in the jungles of Kaladhungi, he was on intimate terms with the big cats. Indeed, he'd had multiple run-ins with the oversized predators, including one encounter that left a lasting mark. As a young boy, not so very long after the death of his father, he was walking alone in the forest when he stumbled right into a large Bengal, peering out at him from a plum bush. The tiger could have easily made a meal of the young Corbett with nothing more than a swipe of its claws, but it did not—it merely watched him inquisitively for a moment with its piercing golden eyes before melting away, back into the forest. It was something he would never forget.

But while Corbett may have known a great deal about tigers— possibly more than any other European in India at that time— unlike most of his British peers, he had little interest in hunting them. He tended to view the apex predators with much the same quotidian blend of awe and respect as the local people he had lived beside in Kaladhungi and Nainital. "A tiger's function in the scheme of things is to help maintain the balance of nature," Corbett would later write, "and if, on rare occasions when driven by dire necessity, he kills a human being or when his natural food has been ruthlessly exterminated by man . . . it is not fair that for these acts a whole species should be branded as being cruel and bloodthirsty." To him, the tiger in India represented "the finest of her fauna," and even at the time of that first crucial conversation with Knowles in 1903, he was already in a good position to know.

Jim Corbett had spent countless nights in his youth hiking through backcountry trails and sleeping in the open; he had dwelt among people, both the Pahari of the lower hills and the Tharu of the plains, who cut grasses and gathered fuel alongside tigers every day. He knew all too well the role that tigers played in keeping the delicate scales of nature from tipping, and he was under no illusions as to their disposition or abilities. Corbett was neither Hindu nor Tharu, yet in his descriptions of the animals, he readily identified something almost omnipotent, perhaps even divine. In short, he understood as only those who live in partnership with the forest truly can that to venture into its borders was to enter the realm of another lord's kingdom. In the jungles of India, the tiger reigned supreme. It was a lesson he had learned with that very first encounter, with that very first pair of blazing golden eyes.

Although such insights may have been relatively rare among British colonists in India at that time, the divinity of tigers—indeed, the *necessity* of tigers—had been a tenet of faith on the subcontinent since ancient times. Consider this text from the *Mahabharata*, the great Sanskrit epic poem of India, first penned—or rather chiseled, as it was initially inscribed on tablets—around 400 B.C.:

> Do not cut down the forest with its tigers and do not banish the tigers from the forest. The tiger perishes without the forest, and the forest perishes without its tigers. Therefore the tigers should stand guard over the forest and the forest should protect all its tigers.

This notion of the tiger as a sort of spiritual guardian is deeply embedded in Hindu cosmology, and it is reiterated and reimagined in various forms throughout Hindu religious texts. To the ancient civilizations of the Indus Valley, the tiger was a potent vehicle utilized by supreme beings. Durga, the warrior goddess charged with driving evil and darkness from the world, is almost universally depicted as riding a tiger as she battles demons. Among those who venerate the forests in India, her protective role takes the form of the spiritual guardian Bonbibi, or Ban Dhevi, who also rides the tiger to defend her jungle kingdom from malevolent forces. Woodcutters and honey collectors have always been especially devoted to her, with the tradition of making puja sacrifices for her protection from wild animals going back centuries.

The bond between humans and tigers was more than symbolic—it was, according to many Indian creation myths, literal as well. A legend originating in the northeastern state of Nagaland tells how man and tiger were both born of the same mother, and emerged as spirits through a pangolin's den. Among the Warli tribes north of Bombay, a wedding cannot be consecrated nor fields planted without first paying tribute to the tiger god Vaghadeva. His blessing is seen as a critical component in any reproductive act, be it farming or childbearing. In all of these cases, tigers appear not as foreign enemies, but as natural allies in keeping the earth both fertile and safe. Their power is something to be respected, and if possible, harnessed, as a means of maintaining balance between the forces of life and death, darkness and light. A world without tigers would be a void, a place of pestilence, a land toppled on its side.

Given the close associations of tigers with powerful, eternal deities, it is not surprising that the symbol of the tiger was enthusiastically adopted by many Indian kings. Royal seals of the Harappan dynasty dating back to 3000 B.C. depict heraldic tigers, and communicate a ritualistic link between the forces of nature and royal decree. Some royal families, like the Tamil Chola dynasty, which ruled much of southern India between 850 and 1014, even went so far as to appropriate the tiger as their official emblem, featuring it on their coins and banners. The Bengal tiger served for many kings as a symbolic emissary ranging across a land that was, at that point, still largely wild and uncultivated. The Chinese chronicler Hieun-Tsang, who embarked on a voyage across India in the seventh century, described an endless country "covered with thick jungle and forest trees with streams flowing round its limits and abounding with a spiritual force." Essentially, a place where human settlement was sparse and tigers flourished. As projections of royal power, tigers represented omnipotence and omnipresence alike. Even as late as the eighteenth century, the Sultan of Mysore was clothing himself completely in tiger stripes, and displaying to his enemies a banner that decreed "The tiger is God."

Of all the dynasties that exercised control over India's many territories, few had a more intimate relationship with tigers than the Mughals. Between 1526 and 1827, this Muslim family of central Asian origin ruled over an empire that at its height covered almost all of modern India. And as both sportsmen and ambitious imperialists, the Mughals recognized early on the ritualistic importance of the royal tiger hunt. Although the hunts had existed since ancient times, it was the Mughals who established and codified the true Indian *shikar*, complete with locally drafted beaters, elaborate camps, and bejeweled elephants. The Mughals enjoyed hunting in all of its forms, from falconry to cheetah coursing, although the *bagh shikar*—the true tiger hunt—was always the means by which they most clearly demonstrated their royal mandate. Elaborate tiger hunts, conducted on a rotating basis at a series of royal hunting reserves throughout their empire, served a function that was diplomatic, militaristic, and even religious. By incorporating local villagers as beaters and headmen as organizers, they were able to forge cooperative alliances with distant vassal states, while at the same time make a clear demonstration of their martial capacity should those alliances ever break down. In the final killing of the tiger—the lord of India's forests—they ritualistically reinforced their own position at the top of the political food chain.

In its early manifestations, the Mughal *bagh shikar* had minimal effect on tiger populations or the habitats in which tigers lived. Held at widely dispersed forests on a rotating schedule, and conducted primarily with bows and spears, these hunts were never intended to deplete the tiger population or rid a region of predators. Indeed, tigers were considered precious royal property—only the king and his vassals had the right to hunt them, and they seldom did so on a scale that threatened their existence as a species. Even a prolific hunter like Jahangir, who served as emperor between 1605 and 1627, reputedly was careful to kill game at a sustainable rate. In the first twelve years of his reign, he was reported to have killed eighty-six lions and tigers—a hefty sum, admittedly. But when one considers the annual average, it comes out to about seven kills a year. As long as the hunting sites were regularly moved, local populations of tigers could be easily replenished. Mughals like Jahangir held the tiger in awe—they placed tremendous value on having tigers in their realm, and they took great pride in their tiger-hunting traditions. Ensuring a healthy population of tigers throughout their territory was more than just a familial whim; it was their responsibility as Indian kings.

The future of India—and of the tiger—would both change forever with the arrival of European merchants. The taste for Asian spices and silks had been acquired by wealthy Europeans during the Middle Ages, although by the fifteenth century, they had grown tired of purchasing from the usurious merchants of the Silk Road and were actively seeking to cut out the middlemen. But whereas Christopher Columbus failed in his endeavor to sail to the Indies—though a fateful failure it was—Vasco da Gama succeeded, and by 1498, the Portuguese mariner had rounded Africa's Cape of Good Hope and made landfall at the spice port of Calicut on India's Malabar Coast. The English weren't far behind, and by 1511, King Henry VIII was already receiving petitions from merchants stating that "The Indies are discovered," and urging the monarch to "bend our endeavours thitherwards" in the pursuit of trade. The first serious incursions wouldn't occur until the latter half of the sixteenth century, though, when a trio of London merchants— Ralph Fitch, William Leeds, and James Story—set sail for the east in 1583, aboard a ship aptly named *Tyger*. Disembarking at Tripoli, they made the rest of the journey by foot, covering three thousand miles on their march to the subcontinent. On the way, they would discover "all sorts of spices and drugs, silk and cloth of silk, ele-phants teeth and much China work, and much sugar." It wasn't all sugar and spice, however—Ralph Fitch found that the dense forests of India were also filled with frightful animals, including tigers. Predators quite unlike any he had ever seen.

Fitch's accounts of India stirred the imaginations of his country-men back home. Playwrights like Christopher Marlowe dreamed of sailors "Lading their ships with gold and precious stones," while authors like John Milton envisioned the fabled riches of "Agra and Lahor of Great Mogul." Though such stories provoked romance in the hearts of England's bards, the nation's merchants saw noth-ing but moneybags. Fitch's stories spurred one especially ambitious group of London traders to approach Queen Elizabeth I with a tempting offer to enrich the nation's coffers. The Queen agreed, and in 1600 she granted the newly established East India Company a royal charter, effectively designating their organization as the pri-mary liaison between Britain and the rich trading ports of the Far East. Initially, their focus was on the Spice Islands, but shifted to India as a result of conflicts with the Dutch. Inroads were made, and by 1612 the East India Company had convinced the Mughal emperor Jahangir—that same legendary hunter and great lover of tigers—to give them permission to trade at the port of Surat. This small mercantilist concession may not have seemed especially portentous at the time, but its effects would be felt for centuries. Without realizing it, Jahangir had helped usher in the British colonial era in India and inadvertently sealed his own dynasty's doom. With Calcutta serving as their colonial capital, the embold-ened British—under the auspices of the East India Company—began to expand their enterprises in India, first in collaboration with their Mughal hosts, and later, in direct defiance of them.

The century that followed saw a gradual consolidation of Company power, and increased tensions with native-born rulers and European rivals. Those tensions would culminate in 1757 with the Battle of Plassey, when the Company's own private army defeated a local Mughal vassal and his French allies on the battlefield, resulting in direct Company control over all of Bengal. After which, any pretense of native Indian rule, at least in the eastern part of the subcontinent, was effectively eliminated. From its base in Bengal, the East India Company continued to expand its hegemony westward and southward over the rest of India throughout the late eighteenth and early nineteenth centuries. Provinces that cooperated were allowed to exist as subordinate "princely states," while those that rebelled were met with ruthless force. The role of the British had substantially transformed. They were no longer foreign merchants looking to make a tidy profit—they had become governors. A small, damp, windswept island a full hemisphere away had come to rule, via the East India Company, a sprawling and complex subcontinent it barely understood. Through both guile and blunt force, the British had usurped the Mughals to achieve power over much of India; what remained to be negotiated, however, was how best to wield it, and maintain it.

The solution came in the form of a carrot-and-stick-style collaboration with India's minor nobility. Specifically, between the local princes, who were generally allowed to keep their fiefdoms in exchange for loyalty and tribute, and British colonial officers, who were charged with gubernatorial oversight. In this way, the local maharajas could continue to rule over their subjects and handle daily administrative affairs, while a British resident backed up by military force oversaw each principality to ensure the Company's interests were advanced. This collaboration in turn created a certain amount of shared cultural space between the Indian and British colonial elites, and the exchange of aristocratic traditions was not uncommon. The children of maharajas learned to play cricket and speak Oxford English, while British ministers developed a fondness for lofty elephant rides and elaborate hunts in jungle reserves. In this exchange, the ritual of the colonial tiger hunt was born.

One of the first published accounts of a colonial tiger hunt comes via a letter penned by Sir John Day, dated April 1784. In the decades that followed his first encounter with tigers, such orchestrated *shikar* expeditions would become commonplace for visiting Britons of means, although at the time of his letter, they were still exceedingly novel and "exotic" affairs. What follows is Day's description of the big event, which he recorded as occurring on the banks of the Ganges, near Chinsurah, in Bengal:

> Matters had been thus judiciously arranged: tents were sent off yesterday, and an encampment formed within a mile and a half of the jungle which was to be the scene of our operations; and in this jungle the thickets of long rank grass and reeds are in many places fifteen feet high. At one o'clock this morning thirty elephants, with the servants, and refreshments of all kinds, were dispatched . . . we mounted our elephants, and proceeded to the jungle.
>
> In our way we met with game of all kinds; hares, antelopes, hog-deer, wild boars, and wild buffaloes; but nothing could divert our attention from the fierce and more glorious game . . .
>
> We had not proceeded five hundred yards beyond the jungle, when we heard a general cry on our left of "Baug, baug, baug!" On hearing this exclamation of "tiger!" we wheeled; and forming a line anew, entered the great jungle . . . on the dis-

charge of the first gun a scene presented itself confessed by all the experienced tiger hunters present to be the finest they had ever seen. Five full grown royal tigers sprung together . . . they all crouched again within new covers within the same jungle, and all were marked. We followed, having formed a line into a crescent, so as to embrace either extremity of the jungle: in the centre was the houdar (or state) elephants, with the marksmen, and the ladies, to comfort and encourage them.

When we had slowly and warily approached the spot where the first tiger lay, he moved not until we were just upon him; when, with a roar that resembled thunder, he rushed upon us. The elephants wheeled off at once . . . They returned, however, after a flight of about fifty yards, and again approaching the spot where the tiger had lodged himself, towards the skirts of the jungle, he once more rushed forth, and springing at the side of an elephant upon which three natives were mounted, at one stroke tore a portion of the pad from under them; and one of the riders, panic struck, fell off. The tiger, however, seeing his enemies in force, returned, slow and indignant, into his shelter; where, the place he lay in being marked, a heavy and well-directed fire was poured in by the principal marksmen; when, pushing in, we saw him in the struggle of death, and growling and foaming he expired.

We then proceeded to seek the others . . . and with a little variation of circumstances, killed them all; the oldest and most ferocious of the family, had, however, early in the conflict, very sensibly quitted the scene of action, and escaped to another part of the country . . .

The chase being over, we returned in triumph to our encampment . . .

As Sir John Day's account demonstrates, many elements of the traditional royal tiger hunt were appropriated by the colonial newcomers—indeed, the techniques the English adopted were not substantially different from those of their Mughal predecessors. They sat perched in the same howdah, atop the same elephant, while the same local shikaris served as guides, and the same villagers were drafted to beat the tigers out of the bushes. The hunt had all the familiar ostentation, all the connotations of military might, and much the same ritualistic importance as that of the Mughals'. The fundamental difference, however, was what the ritual signified. The tiger held a different implication for the British than for the Mughals and other Indian rulers.

Under the former, the ritualistic slaying of the animal took on a new meaning—and the tiger itself would become a symbol of local defiance. Consider Tipu's Tiger, a peculiar, life-sized mechanical curiosity that once belonged to Tipu Sultan, the ruler of the rebellious kingdom of Mysore. Unlike many of his neighbors, Tipu Sultan refused to capitulate to the East India Company. Fiercely proud—indeed, the Sultan liked to call himself the "Tiger of Mysore" and used the animal as his official emblem—he would not accept British encroachments into his territory, or form alliances with them of any kind. It was a courageous strategy, although not a successful one. The Sultan was defeated in the Fourth Anglo–Mysore War, and in the end British troops stormed his palace, discovering a bizarre and most unsettling sight, as recorded in this East India Company account published in 1799:

> In a room appropriated for musical instruments . . . was found
> an article which merits particular notice, as another proof of
> the deep hate, and extreme loathing of Tippoo Saib towards

the English: this was a most curious piece of mechanism as large as life, representing a Royal Tiger in the act of devouring a prostrate European officer: within the body of the animal was a row of keys of natural notes . . . intended to resemble the cries of a person in distress intermixed with the horrible roar of the Tiger: the machinery was so contrived, that while this infernal music continued to play, the hand of the European victim was often lifted up, and the head convulsively thrown back to express his helpless and deplorable situation. The whole of this machine, formed of wood, was executed under the immediate orders, and direction of Tippoo Sultaun, whose custom it was in the afternoon to amuse himself with this miserable emblematic triumph . . .

The mechanical device with its haunting organ depicted an actual event—Hugh Munro, the only son of a British general who had defeated Tipu's own father in a previous Mysore rebellion, had been killed by a tiger while hunting with friends on the Bay of Bengal in 1792. The device's commemorative aspects, however, were overshadowed by its symbolic message. And that larger meaning was hardly lost on the British soldiers who had stumbled upon it. The triumphant tiger represented all the innate, unyielding ferocity of a native-born Indian king, exercising his divine right to repel a foreign invader, while the wailing British officer represented—well, a wailing British officer. Tipu Sultan had used his emblem, the tiger, as a projection of his own defiance against the British, and the British were more than happy to bend that analogy to fit into their own worldview. Whereas the kings of India had regarded the tiger as a powerful yet benevolent manifestation of their own royal mandate, the British viewed the very same animal with suspicion and disdain. And while the hunting of tigers had once been an affirmation of identity for India's ruling class, it became under the British a sort of ritualized reenactment of the colony's subjugation—the slaying of the rebel Tipu Sultan performed all over again, albeit on a smaller scale. To kill a tiger was to vanquish all that was deemed alien and dangerous in India; the act itself made the country safer, and a little more like England. The existence of tigers in the wild was viewed, both symbolically and literally, as a direct challenge to British hegemony. Overcoming that challenge was an act of conquest—of colonization—and it was very much encouraged by the colonial government.

Viewed through this new colonial British lens, the goal of tiger hunting changed as well. To a Mughal king or a local maharaja, the idea of actually exterminating tigers would have been preposterous. To hunt tigers was their divine right, a sacred ritual of affirmation; to remove *all* the tigers from the Indian landscape would have imperiled their own identity. To the British, however, the extermination of tigers became synonymous with progress, with "civilization." They were an obstacle to modernity that needed to be removed. For British governors and the military officers who carried out their orders, the tiger was, as the journal *Tiger-shooting in India* put it, a "cunning, silent, savage enemy" that committed "fearful ravages" against those trying their best to "civilize" what was, in their own Eurocentric eyes, a "primitive" land. How could an ambitious colonist establish a tea plantation, or a cattle station, or even a decent school, with bloodthirsty five-hundred-pound predators lurking in the shadows? To the British in India, this was a rhetorical question.

Ample evidence of these attitudes can be found in the many

field manuals on shooting and hunting published by colonists at the time. In one titled *Oriental Field Sports,* printed in 1807, Captain Thomas Williamson expounds on the benefits of tiger eradication:

> Of such importance has the search for tigers, and their consequent destruction, proved in some parts of Bengal, that large tracts of country in a manner depopulated by their ravages, or by the apprehensions to which the proximity of such a scourge naturally must give birth, have, by persevering exertion, been freed from their devastations; and, in lieu of being over-run with long grass and brambles, have become remarkable for their state of cultivation to which they have been brought. Perhaps no part of the country exhibits a more complete corroboration of this fact than the Cossimbazar Island; which, though not exempt from the evil, has changed from a state of wilderness to a rich display of agriculture. A few patches of cover yet remain; however, they cannot fail to speedily be annihilated, when perhaps a tiger may be as great a rarity, as formerly it was an incessant object of terror.

In just one short passage, the good captain essentially sums up the entirety of early British attitudes toward tigers, and toward India as a whole. He sees the forests and grasslands merely as places to be cleared and put to the plow, and the predators that inhabit them as actually "evil" in their unwillingness to be subjugated or easily removed. Colonialism in a nutshell.

That isn't to say, however, that colonial attitudes toward tigers were *completely* unfounded. Although generally shy and reclusive, tigers did occasionally kill livestock, and sometimes did indeed kill people. Bullet wounds weren't the only infirmity that could lead to man-eating—old age, worn-down teeth, and even porcupine quills occasionally turned otherwise normal tigers into man-eaters, and human–tiger conflict was not unheard of in British India. In 1769, tigers in the forests around Bhiwapur were said to have claimed over four hundred victims, causing an entire village to be abandoned. And more than fifty years after poor Hugh Munro, the British general's son, succumbed to stripes on the Bay of Bengal in 1792, some seven hundred people a year were still being killed by tigers in the state of Bengal alone. Obviously, tigers weren't harmless—they were apex predators, after all, endowed with awesome strength and terrific abilities. And humans were, at the end of the day, edible meat. But when one considers their actual menace compared to other animals, it becomes obvious that the danger they posed was greatly exaggerated. The annual government gazettes from the period clearly show that other animals presented a significantly greater threat to humans than tigers. Poisonous snakes, for example, consistently killed twenty times as many people in India as tigers throughout the nineteenth and early twentieth centuries, and many other "wild beasts" such as wolves, bears, leopards, and rhinos rivaled and regularly surpassed tigers in terms of yearly victims. Out of the annual total of between 20,000 and 25,000 lives that were usually claimed by wild animals in India each year during the colonial period, tiger victims generally numbered somewhere between 800 and 1,000. And while 1,000 deaths a year due to tigers may seem considerable, out of a total population already approaching 300 million by the late nineteenth century, it's still a very low mortality rate. When one considers that nearly five hundred people die each year in America—a country with a population roughly comparable to India's in the nineteenth century—simply from falling out of bed, one realizes that except for a few isolated occasions and regions, tigers would have been the least of the average Indian villager's worries. And even in the rare instance when tigers did present a realistic threat, the people were not, despite their portrayal by colonial rulers, totally helpless. Many rural communities did have their own methods for dealing with tigers—after all, they had been living alongside them for hundreds if not thousands of years. In 1815, for example, a judge in Madras reported how seven hundred villagers "formed a Circle round [a] Tyger" that had been terrorizing their lands and finished it off with spears. There are similar accounts of villages banding together to use nets and even poisoned arrows to get rid of problematic tigers—something the British, who generally lived in cities and bungalows far removed from any actual threat of tiger predation, had the audacity to critique as unsporting. Apparently they saw more valor in butchering tigers at scale, using high-powered rifles from an elephant's back.

The tigers' tragic fate under colonialism also produced unintended human consequences. Although it may seem counterintuitive, the increased hunting of tigers can and often does translate directly into more human deaths by tiger. Obviously, there is a point at which so many tigers have been killed, there are simply none left to strike back—and this was precisely what would happen in India in the second half of the twentieth century, when tigers were brought to the brink of extinction. But there was also a point, earlier in the process of decimation, during which large numbers of humans were entering the woods and actively seeking to interact with large numbers of tigers. A simple increase in human–tiger interactions inevitably results in an increase in lethal encounters, and as we well know, simply having a rifle is not always an effective defense against a full-on tiger assault.

Valmik Thapar took note of this correlation in India's Ranthambore Park in the 1970s—after tiger hunting was totally banned by the government, the number of tiger attacks fell steeply. A more compelling, and exact inverse, case study for our purposes is one first documented in the Russian Far East in the 1990s. Following the collapse of the Soviet Union, desperate economic circumstances forced many locals to increase their visits to the forest, both for the hunting of food and the poaching of Amur tigers. The end result was an unprecedented and drastic rise in tiger attacks over the next two decades—many of which were retaliatory attacks directed at hunters who had fired upon tigers. In one of the more incredible cases, recorded in 2004, a Russian poacher spotlight hunting from an all-terrain vehicle discovered a tiger and took a shot—the tiger immediately charged the vehicle, leapt on top of it, and fatally mauled the unwise hunter. And even when tigers did not directly retaliate against trigger-happy hunters, they sometimes got their "revenge" at a later date. Wounded, bullet-riddled tigers became relatively commonplace in the frigid taiga forests of the Russian Far East, and unable to hunt their natural prey, they turned their attentions toward the slow-footed creatures who had injured them in the first place. A tiger in Russia's Primorye Province that had been disabled by a poacher's bullet in 1997 seemed, based on journalist John Vaillant's excellent reporting, possessed by a personal vendetta: it found its human attacker's house, waited patiently for him in the snow, and proceeded to kill and eat him upon his return. The wounded tiger would go on to kill and eat one more victim, and attack an entire group of local officials who were seeking to end its spree. When the tiger was finally dispatched with assault rifles and its body studied, it was found to have absorbed over the course of its life an extraordinary amount of lead—in addition to the lethal shots that had just been fired, the tiger had a festering flesh wound on its left forepaw, the buckshot from two separate volleys lodged in its right leg, a steel bullet from another rifle stuck deep in its flesh, and a widely dispersed peppering of bird shot throughout its body. And whether it was from a primitive form of feline revenge, or simple hunger, it was no surprise to anyone involved that a cat treated as such by humans might turn that aggression back upon them. Just like the Champawat, this Amur tiger was not born a man-eater—it was turned into one.

Essentially, what occurred in the Primorye Province in the 1990s was happening on a larger scale across India throughout the nineteenth century. Tiger hunting, a ritualized activity that had once been the sole domain of kings, instead became a common form of diversion for colonials. Almost every British officer desired to "bag" a tiger. The Indian *shikar* became the equivalent of the African safari for colonial visitors, and no tour of the subcontinent was considered complete without one. In terms of tiger pelts, single European hunters accomplished over just a few bloody hunts what had taken Mughal emperors like Jahangir a lifetime to achieve. The slaughter of tigers had begun, with the arrival of the British, on an unprecedented scale. In the year 1872 alone, a colonist named Gordon Cummings killed a total of seventy-three tigers along the Narmada River Valley. Another named William Rice shot 158 tigers in Rajasthan over the course of four years. The infamous Colonel Nightingale could claim at least three hundred tigers to his name, shot across the Hyderabad region in just a few years, and George Yule, a member of the Bengal Civil Service who hunted prolifically in the mid-nineteenth century, stopped counting after he had single-handedly bagged four hundred tigers. And these are just a few examples—the field manuals and sporting journals of the day are brimming with similar exploits, full of braggadocio and mocking contempt for the Indian tiger. Not surprisingly, with every officer or high-level administrator doing his best to bring home a tiger skin, many a poorly aimed shot was fired—which would have left the country crawling with wounded, aggressive, and starving tigers ready to sink their teeth into human necks.

Perhaps even more of a catalyst for human–tiger conflict, however, was the practice of bounty hunting. Colonial elites may have hunted tigers for sport, but many more of their subjects hunted, poisoned, and trapped them for money—a direct result of the monetary rewards the colonial government placed on all sorts of "vermin," which tigers were officially categorized as for much of the nineteenth century. Prior to the East India Company's assumption of power in Bengal in 1757, bounties placed upon animals had been virtually unknown. Within just a few decades of British conquest, however, bounties on any species deemed unwelcome or unnecessary were commonplace. And tigers, at the time, were most unwelcome indeed. Indian shikaris, who tended to be lower-caste Hindus of limited means, were recruited from every province to help eradicate the tigers of India. Driven by a pursuit for liquid currency to survive in the colonial system, they generally complied. A tiger could fetch a considerable sum for the average colonial subject, as described in this account from *Oriental Field Sports,* published in 1807:

The death of a tiger is a matter of too much importance to be treated with indifference. The Honourable East India Company, with the view to prevent interruption to the common courses of business, and to remove any obstacle to general and safe communications, bestow a donation of ten rupees, equal to twenty-five shillings, for every tiger killed within their provinces. The Europeans at the several stations situated where the depredations of tigers are frequent, generally double the reward. Besides the above allurement, the sale of the skin, claws, &c. often amounts to nearly as much more; forming in the aggregate a sum which, in a country where an ordinary person may board, lodge, and clothe himself comfortably for ten shillings monthly, may be considered quite a fortune. Under such a forcible temptation, the shecarrie repairs to the place; and being guided by the peasants best acquainted with the jungle wherein the tiger is concealed, he proceeds to search for the carcase.

Such bounties would not only persist but grow throughout the course of the nineteenth century, and even well into the twentieth. Between 1875 and 1925, over eighty thousand tigers were slaughtered for government bounties—and those were just the kills that were officially recorded. It's likely that many more tigers were taken, be it for sport or to protect livestock or simply to sell their body parts.

With nearly anyone who could get their hands on a gun taking potshots at tigers, it's no wonder that the incidence of wounded, aggressive animals increased. The tiger population may have begun falling, but incidents of tiger attacks would only rise, as amateurs ventured into the forest in greater numbers, and more tigers were left limping through the undergrowth, riddled with bullets and injured by traps.

A surge in tiger hunting among Europeans and Indians alike

may have been a contributing factor to the rise of man-eaters in India, but it wasn't the primary culprit—not by a long shot. What truly drove hungry tigers out of the jungles and into the villages was far less exciting, but significantly more deadly. When Britain took India as a colonial possession, it did far more than simply introduce a new power structure or administrative policy—it effectively turned the entirety of its foreign possessions into an engine of revenue, which meant exploitation of natural resources on a massive, multifaceted scale. And in few places would it prove as profound—or as lethal—as in Kumaon.

The place where the Champawat would truly make its mark.

Unlike the British territories in Bengal, the lands composing the Kumaon division had never been under the direct control of the Muslim Mughals. The various princely states that made up the territory were variously consolidated and broken up again by a series of shuffling dynasties over the centuries—the last of which was the Hindu Shah dynasty of Nepal, who conquered the kingdom of Kumaon as part of their consolidation of the Nepalese state in 1791. The cultural differences were all but negligible between Kumaon and their holdings in western Nepal, with the same Tharu people inhabiting the lowland *terai,* and similar Pahari tribes to be found in the middle hills. In short, in terms of territorial acquisition, it seemed to make sense. The Nepalese expansion across the Sharda River was short-lived, however—the British East India Company, which had not forgotten the disastrous campaign of Captain Kinloch in the *terai* in 1767, finally got their revenge and some much-coveted Gorkha territory with the Anglo–Nepalese War of 1814. The British were more savvy this time around, and after a series of stinging defeats, the Nepalese finally agreed to hand over their lucrative trade routes with Tibet, as well as one-third of their total territory—including Kumaon. In one fell swoop, the British had added a substantial chunk of northern real estate to their rapidly expanding holdings in India. Following the Shah dynasty's capitulation, Kumaon was incorporated into British Bengal. In 1835, in recognition of the difficulties of governing such a vast territory, however, Kumaon became part of a new colonial state called the Northwest Provinces. Initially, the people of Kumaon had welcomed the British, as they saw them as a means of shedding Gorkha rule—which they certainly were. The people of Kumoan were less enthused, however, when it became clear that the British had no intentions of leaving. The new colonial government of the East India Company quickly began exploiting the region and its resources in ways the Gorkhas had never dreamed of. While there was a brief but bloody Kumaoni uprising as part of the larger "mutiny" of 1857, the rebellion was crushed by British forces, and a new draconian system of direct government oversight was put into practice. The English Crown had won full control of the region, and it proceeded to do what it did best: squeeze the land for all it was worth. Which meant depleting the natural resources of Kumaon on an epic scale, and making its farmland as productive as possible. Making it "civilized," as it were.

Under the new British Raj, the formerly lax agricultural oversight of the Gorkhas was overhauled. Spearheaded by Sir Henry Ramsay, who served as the commissioner of Kumaon from 1856 until 1884, this agrarian revolution meant a drastic increase in the amount of land put to the plow, and the number of people toiling in the fields. The agricultural population grew by 13 percent in 20 years, and the amount of cultivable land increased by 50 percent in the last 4 decades of the nineteenth century. By encouraging the settlement of land and crop rotation based on the Kharif, or monsoon, the government was able to take what had been a ragged, sparsely populated collection of bhabar hills and marshy *terai* valleys and turn it into something of a regional breadbasket, producing rice, wheat, barley, and mandua to supply a growing colonial population and further enrich the government's coffers with tax revenues. Terraced slopes and alluvial valleys provided much of the bounty; however, in rocky areas where the soil was poor, canals supplied irrigation, and in lowland regions where malaria was prevalent, tenant farmers from the hills would sow crops when the season allowed, then return to their high-altitude homes to wait out the mosquitoes. This maximization of arable farmland continued into the twentieth century, with jaw-dropping increases in grain production. From 1901 to 1911, the number of acres cultivated for rice in Kumaon rose from 70,239 to 400,623—a leap that was made possible thanks largely to irrigation, which increased for all farming from 169,602 irrigated acres to 357,419 acres in that same time span. These new irrigated lands were fed by a total of some two hundred miles of canals that in turn allowed southern rice strains like basmati and *hansraj* to be grown in the dryer, northern climate. Wheat, barley, and corn production also saw significant growth in this time, while the total amount of uncultivated land—timber excluded—decreased from 1,608,119 acres to 900,280 acres during that decade.

The march toward productivity that Commissioner Ramsay initiated extended beyond agriculture, and included the management of forests as well—not as nature reserves, or even as game reserves, but primarily as a source for the vast amount of timber needed to build the colony's burgeoning railroad infrastructure.

As in the American West, the railroad opened up the frontier and allowed new settlements to prosper, albeit at the expense of the colony's timber supply. Ironically, Jim Corbett himself was employed by the Bengal and North Western Railway, which was devouring the very forests he loved above all else. The industry's appetite for timber was due largely to a tremendous increase in the production of railway sleepers, and the sturdy sal trees, so crucial to the ecosystem of the region, fit the bill nicely, as they were uniquely suited for India's climate. Yet by the latter decades of the nineteenth century, the sal forests of northern India were in serious decline due to over-logging. A British nautical journal published in 1875 reports:

> The sal forests of Upper India might be shown (according to the report of Mr. Webber, of the Indian Forest Department) to be even in worse plight through reckless cutting and utter neglect. There were in 1830 probably 4,000 square miles of purely sal forests along the foot of the Himalayas, beside those in Central India, available to Government . . . yet now, the East Indian Railway has been obliged to import pine sleepers from Norway, sal being scarcely procurable.

Realizing the sal forests were all but destroyed, the savvy colonial government again decided something needed to be done to protect its crucial timber supply. Under the Indian Forest Act of 1878, special reserve forests were set aside for the colony, with the aim not of saving India's wildlife or protecting its ecosystems, but of preserving its viability as a source of construction lumber. As commissioner, Sir Henry Ramsay was a vigorous proponent of such measures in Kumaon, and he is often credited with "protecting" the district's forests. This is true in that he essentially kicked many indigenous Tharu and Pahari people out of them—people who had lived in a relatively harmonious, even symbiotic relationship with the forest for centuries—to ensure that the colonial government had a total monopoly on the harvesting of the wood that remained. His policies of forest management, if one can call them that, would live on well after his tenure was over. In the same ten-year period between 1901 and 1911 in which the productivity of rice and other crops exploded, the total area of forests set aside for the harvesting of timber surged tremendously as well, from 174,142 acres to 3,781,503 acres. To the neighboring Tharu and assorted Kumaoni hill tribes that relied on the forests for their livelihood, this severance from their ancestral birthright was tantamount to the extirpation of the American buffalo to the Plains Indians; it totally altered their way of life, and threatened their continued survival. The hunting and gathering, the grass cutting, and the animal grazing that had defined their sylvan existence were all at once severely restricted. The tribes did not take such attacks sitting down, and although Corbett elides over it in his writings, anger over forestry practices and even full-scale revolts and wildfires lit in protest were not uncommon in the first decades of the twentieth century. Essentially, the few sal forests of the *terai* and middle hills that remained intact were transformed into fenced-in tree farms, places totally off-limits to anyone but loggers and officials, where deer and other ungulates were seen as a threat to young trees, and where tigers—well, unless they were at the end of a sportsman's rifle, tigers had virtually no use at all.

And even the assorted deer species that were not driven off through habitat destruction and poaching often fell victim to fresh waves of disease. Rinderpest, in particular, was common in domesticated cattle and goats, and with the influx of livestock and cattle

stations that occurred in Kumaon in the latter decades of the nineteenth century, it is not surprising that the disease drastically affected the ungulate populations in nearby forests. A doctor's 1894
account of rinderpest, penned in the Kumaoni hills, takes note:

> The malady is very commonly observed in the buffalo, amongst
> which it is very fatal. The yak also suffers from a very severe
> form of the disease, as do all his crosses with the cow, and it is
> very fatal in them . . . Antelope also suffer from the disease, as
> do cervidae generally. The gooral and kakar, or barking deer,
> have been seen dying in large numbers from the disease in the
> Himalayas.

In fact, a severe outbreak of rinderpest was recorded in Kumaon between 1899 and 1900 by an animal specialist engaged in
inoculation—he attributed its spread to sheep and goats from the
lowlands at the "foot of the hills," which according to him, "is
seldom free from the disease." The foot of the hills, incidentally,
would have been precisely where a normal tiger would have sought
its prey. Similar plagues occurred repeatedly across northern India
in the late 1800s and early 1900s, involving rinderpest, hoof-and-
mouth disease, cholera, even anthrax. In some parts of the region,
they wiped out the wild deer populations completely.

With the profound changes taking place in Kumaon's agriculture and forestry—as well as parallel policies instituted around the
same time by the Ranas across the border in Nepal—the emergence of a tiger like the Champawat was not merely possible, it
was, perhaps, inevitable. The grasslands where chital deer thrived
were being put to the plow; the forests where sambar and gaur
made their home were being logged at an unprecedented rate and

drained of their biodiversity; the local people, who had lived sustainably alongside these places for millennia, were suddenly deprived of much of their livelihood and forced to sneak into the forest at night like bandits, stealing animal fodder and poaching game. Of course something was bound to give, not just with tigers, but with predators in general. And indeed, it did.

A government report from the United Provinces covering wild animals in 1907 and 1908 attributed a drastic increase in predatory leopard attacks in Almora to "wholesale destruction of game such as sambhar, gural and kakar," all of which resulted in "a serious diminution of the natural food supply of tigers and leopards." Similarly, a sudden rise in wolf attacks recorded in nearby Allahabad in 1906—a local pack carried off eighty-six children that year, compared to nineteen the previous one—was attributed to "the growing scarcity of game in the district and to the consequent laps into bad habits of individual wolves." And according to that same report, the total number of fatal tiger attacks in Kumaon—just those that were recorded by the government, mind you, for there were surely many more—jumped by 500 percent in that year alone. This last statistic was almost certainly caused by the Champawat, a tiger whose own bad habits put the wolves of Allahabad to shame.

But what other recourse would such a tiger have had? With its bullet-damaged canines, it would have been unable to hunt most of its natural prey. The smaller deer it might have killed with its paws would have been almost too scarce to hunt. The tall grasses and deep jungles where it should have lived a solitary, near-invisible existence were vanishing by the day. The forest-dwelling Tharu who revered tigers were being supplanted by a population of colonials and migrants who viewed big cats simply as a menace to themselves and their livestock. Ask any person with actual tiger experience, and they'll tell you: although tigers are not by nature aggressive toward humans, when cornered or in peril, they will attack. And in the hills and valleys of Kumaon at the dawn of the twentieth century, the Champawat Tiger had nowhere left to run. It was both cornered and in peril.

Nevertheless, it would be mistaken to imply that empathy and understanding of this fact was completely beyond the scope of the colonial mind. Granted, the majority of those in the government or civil service did view tigers as a valuable trophy at best, or at worst, a bloodthirsty pest. But not every British colonist living in India reduced them to the status of game or vermin, and Corbett's was not the only voice that spoke highly in their favor. In a surprisingly prescient and sensitive editorial published in 1908 titled "A Wronged Animal: Justice for the Tiger," *The Times of India* made the case—no doubt scandalous at the time—that the tiger was "not only a harmless but a useful member of Indian society," and that the colonial government was "oblivious to the really valuable services the vast majority of tigers render to the Indian cultivator— services unsullied by wrong doing or even intimidation." This is an obvious reference to the fact that tigers traditionally had kept wild deer and pig populations at levels that prevented them feeding on the crops of villagers. Also noted was that, despite the occasional cattle kill by a tiger, the average farmer, when "weighing the services rendered him in the uneven warfare he wages in defense of the crops which stand between him and starvation, he prefers the tiger alive to the tiger dead." This sentiment, while revolutionary for a European colonist, is again very much an echo of what many Tharu and some Pahari peoples, who had lived alongside tigers since time immemorial, knew all along: the predators helped to maintain a sustainable balance for those whose lives depended upon the forest, and the loss of the occasional cow or goat was a small price to pay for a healthy harvest and an adequate food supply. And this balance persists in the modern day, in the pockets of Kumaon where traditional agriculture survives. Among the small farmers I spoke to, both Tharu and Pahari, all expressed an appreciation for predatory cats for protecting their fields from boar and deer. In places where tigers and leopards were scarce, farmers had to stay up all night guarding their fields in a cramped machan—a grueling task no one was keen on. In the nineteenth and early twentieth centuries, however, this sort of small-scale, subsistence agriculture was already beginning to disappear across wide swathes of Kumaon, and the ingrained cultural appreciation for the tiger was vanishing right alongside it. Why would a colonial government bent on populating a sparsely inhabited region and pushing economic productivity through the roof care about the delicate balance among tigers, wild ungulates, and forest-dwelling locals, when all of the above could be wiped off the face of the earth and replaced with field after field of cultivable and taxable land?

In light of such facts, there is an intriguing historical comparison that all but begs to be made. Essentially, the powers that be were seeking to do in India over the course of a few decades what had already been accomplished in Britain over several centuries: the transformation of a forested land rich in both prey and predators into a productive, pastoral agro-scape utterly devoid of both. The British mandate to "tame" the wilderness had been acted out on its own lands long before it was shipped abroad to any of its colonies. When the Anglo-Saxons arrived in Britain in the fifth century, they alighted upon an island abounding with lynx, bears, and wolves. The first two had been hunted to extinction by the early Middle Ages, and the latter, deemed a threat to sheep and humans alike, were totally eradicated by the seventeenth century. And the forests of Britain, which in ancient times covered roughly half of the island and were widely used as a sustainable resource by local Celtic and Anglo-Saxon tribes, had been trimmed down by the time of the Norman conquest to the point where they covered only around 15 percent of the island. Of that remnant, much was converted to royal reserves, where commoners were prevented from engaging in traditional hunting and woodcutting practices, while noble elites—as any fan of Robin Hood can tell you—essentially exploited the forests as they saw fit. The comparison between thirteenth-century Sherwood Forest and a nineteenth-century Kumaoni forest may seem a stretch, but there are surprising and undeniable similarities: a foreign colonial government imposing forestry restrictions on its native-born subjects, rising taxes that encourage agricultural and pastoral expansion, and mounting bounties put upon the apex predators that stood in their way. And in an almost eerie parallel, King Edward I, who reigned from 1272 until 1307, ordered the total extermination of all wolves in his kingdom, and employed a hunter named Peter *Corbet* to accomplish this where the wolves proved most dangerous. And dangerous, back then, they certainly were. Although wolf attacks would be virtually unknown later in North America—a continent with comparatively ample room and prey—the wolves in Britain, under very similar stresses as the latter-day tigers of northern India, would become just as cornered and just as deadly. In Scotland, the last outpost of wild wolves in Britain, the few specimens of *Canis lupus* that remained by the late sixteenth century were so aggressive toward humans, special shelters called *spittals* were erected along the highways to protect travelers from attack. In the Highland county of Sutherland, the wolves grew so desperate and bloodthirsty, they even took to digging up corpses from graves—a state of affairs that eventually forced the inhabitants of Eddrachillis to bury their dead on nearby Handa Island, where the wolves could not reach them. Memories of those dark days would persist well into the nineteenth century, as demonstrated in these selected verses from the ballad "The Wolf of Ederachillis," published in 1860:

On Ederachillis' shore
The grey wolf lies in wait,—
Woe to the broken door,
Woe to the loosened gate,
And the groping wretch whom sleety fogs
On the trackless moor belate.

The lean and hungry wolf,
With his fangs so sharp and white,
His starveling body pinched
By the frost of a northern night,
And his pitiless eyes that scare the dark
With their green and threatening light.

He climeth the guarding dyke,
He leapeth the hurdle bars,
He steals the sheep from the pen,
And the fish from the boat-house spars;
And he digs the dead from out of the sod,
And gnaws them under the stars.

Thus every grave we dug
The hungry wolf uptore,
And every morn the sod
Was strewn with bones and gore;
Our mother earth had denied us rest
On Ederachillis' shore.

A haunting reminder of what large predators are capable of when they have been pushed to the edge and have nowhere left to run, be it in the far north of Britain like the Wolf of Eddrachillis, or the far north of India like the Tiger of Champawat.

But just like those wolves in the Scottish Highlands, it seemed the Champawat's end was nigh as well. In early March 1907, news broke that a British soldier by the name of Edward Harold Wildblood, touring India while on leave from the Leinster Regiment in Mauritius, had shot the man-eater on *shikar* in the eastern hills of Kumaon. The report was ballyhooed in papers and social clubs across India, and Wildblood was even offered a reward of two hundred rupees for dispatching the most sought-after tiger in all of the empire. It was the sort of story colonial society relished—a dapper, square-jawed British officer in His Majesty's Army, vanquishing an unruly tiger while on holiday. *Oh, what jolly good sport it must have been.*

Or so they thought, anyway.

CHAPTER 5

THE HUNT BEGINS

For four years, Jim Corbett neither heard nor thought about the tiger his friend Eddie Knowles had mentioned on their trip to Malani. He returned to his railway station job at the distant Bengalese river outpost of Mokameh Ghat, where he had risen into the management ranks—gone were the days when he lived in the malarial jungle camps of the Bengal and North Western Railway, helping armies of Indian laborers clear out timber for fuel. He had a desk job now, and authority, although the conditions were still rustic and the pay anything but enviable. Days were often eighteen hours long, and the headaches that came with trying to unload trains and get their goods across a monsoon-swollen Ganges were considerable. And it's hard to imagine that the mingling of mud-flats and coal smoke and colonial disarray didn't create a Conrad-worthy nightmare at times, for Corbett in particular. The heat, the smells, the seemingly endless barrage of crates and people, all would have been oppressive to someone most at home in the quiet hills of Kumaon. Still, it was a decent job for a European of his station—opportunities for the native-born in the colony, particularly those of Irish descent, were few and far between—and despite the hurdles of distance and isolation, he was more or less content with his work. He saw it as a duty of sorts, one that he executed with as much cheer as he could muster, rendering a service that, while stressful, even backbreaking at times, he truly believed was good for the colony. His familiarity with both Indian and British customs made him a natural fit for the position, and while his employers kept him perennially busy, he was still able to sneak in the occasional visit back to the towns and forests of his beloved Kumaon, more than six hundred miles away. By scrimping and saving, he had also managed to cobble together enough funds to invest in a small hardware store in the hill town of Nainital, where he had spent his summers as a boy and where his mother and sisters still lived. His trips back home gave him a chance to check up on the business, as well as relax and reconnect with old friends.

It was on one such visit, toward the end of April 1907, that Jim received a guest in the form of Nainital's deputy commissioner, Charles Henry Berthoud. One can imagine the meeting as quite cordial. After all, the two young men were longtime acquaintances, and the commissioner was, in Corbett's own words, a man "loved and respected by all who knew him." London-born and Oxford-educated, Berthoud relied on Corbett's local knowledge when it came to dealing with problems for which his own experience was limited. Even after a decade in India, he was still an outsider compared to his old friend Jim, and being roughly the same age—both men were now in their thirties—it would have been natural for Berthoud to approach Corbett when he was back in town, with any challenges that were beyond his ken.

One of which, on this occasion, happened to involve a tiger. India was, even in 1907, a country teeming with tigers—perhaps as many as 100,000 of them were to be found in Asia at the turn

of the century—although the numbers were dropping fast. "A tiger" could have been a reference to any one of the cats that were
still hunted by officers for sport, and dragged in for bounties by
shikaris on a daily basis. It quickly became clear, however, which
tiger Berthoud was referring to. *The* tiger. The same man-eater that
Knowles had mentioned four years prior. Not only had his brother-
in-law B. A. Rebsch, the expert tiger hunter, failed to stop it then,
but apparently Wildblood had botched his more recent attempt as
well. The latter had, in fact, shot the wrong cat, a revelation that
had been made public only a few days before, in an article from
The Times of India published on April 15, 1907. The journal ac-
knowledged that the man-eater was still on the loose, admitting
that "the depredations have not ceased" and that a "plea for more
drastic action in dealing with man-eating tigers" was needed—all
of which was no doubt embarrassing for government officials like
Berthoud, who had applauded its demise only weeks before. No,
the tiger was still out there, and killing at its usual horrific rate.
In the years it had free rein in Kumaon, it had raised its human
tally to the unthinkable sum of 434 victims.

To Corbett, the news came as a shock. Perhaps not the failure
of Wildblood to bag the man-eater in question—after all, what
would a soldier on holiday from Mauritius know about hunting
tigers? But the idea that Knowles's brother-in-law had also failed
at the undertaking must have been unsettling. Corbett knew of
Rebsch, by reputation alone. He was the same expert hunter that
Inspector General of Forests Sir Sainthill Eardley-Wilmot would
go on to claim "has shot as much game as any man in India."
Rebsch had even survived a surprise bear attack, suffering a severe
mauling before defeating the beast single-handedly. How was it

possible that the most fearless hunter in all of Kumaon had tried and failed to stop this tiger? And in pondering that question, the true question at hand presented itself.

Berthoud was asking him to go after the tiger. Corbett was their last hope.

॥॥॥॥॥॥॥॥॥॥॥॥॥॥॥॥॥

It's impossible to know precisely how Jim Corbett felt upon receiving the request to kill the tiger. In his written accounts of the hunt, he doesn't go into much detail about his feelings at that point, or his apprehensions. It seems he viewed the request, much like his work for the railroad, as his duty—an unpleasant task that was crucial for the colony. And perhaps—*perhaps*—he also had some inkling of how dispatching a notorious man-eater could help his career down the line. After all, if he was investing in businesses in Nainital, it seems at least plausible that he was already aspiring toward a more entrepreneurial life than the rails could give him.

Either way, Corbett accepted the charge, albeit with two stipulations: first, that all existing bounties on the tiger be withdrawn, and second, that all the hunters and soldiers who were already in pursuit of the tiger be called in. Corbett would later attribute these conditions to an aversion to "being classed as a reward hunter," and to the risk of "being accidentally shot." Both are fair points, and Corbett was known throughout his life to have a fixation with being wounded by other hunters—a relic perhaps of his younger days, when crude rifles were unreliable and it was easy to be mistaken for game by local poachers. But the preconditions are also telling in what they reveal about the ambiguities of his own identity. It would have been widely known in Kumaon at the time that tiger hunting was the bailiwick of two very different factions of colonial society. For the high-born British official, it was a form of aristo-cratic sport. For the Indian shikari, it was a somewhat desperate means of collecting a bounty. Jim Corbett's two conditions were in a way a declaration of his own identity separate from the two. As a native-born colonist of Irish descent, who had grown up thor-oughly intertwined with the local Indian population, he occupied an interstitial space that separated the two worlds—he was proud of that identity, ambiguous as it was, and he wished to ensure he was not mistaken for either a pompous aristocrat bagging tigers for fun, or as a desperate poacher looking to make a few rupees. Jim Corbett wanted to make his purpose clear: he would hunt the tiger on his own terms, and strictly—or at least primarily—out of a sense of duty.

Yet even the notion of "duty" would have been somewhat fraught for someone like Corbett, a colonist whose sympathies and allegiances were sometimes at odds with each other. Colonial atti-tudes had shifted since the early days of the Raj, and this evolution of the British colonial mentality had transformed the relationship between the colonial state and its subjects. The Indian Rebellion of 1857 had been a tragic and bloody affair for all involved, particu-larly in Kumaon. Jim Corbett's own uncle, Thomas Bartholomew, had been tied to a tree and burned alive during the course of the uprising. Violent as it was, however, the insurrection was crushed—it was the last major armed challenge to their authority that the British would face. In its aftermath, control was transferred from the East India Company directly to the British Crown, and the provinces of India were ruled as an integrated part of the Brit-ish Empire. The Indian population was generally prohibited from keeping weapons on hand or gathering without permission, and gradually, in the new "pacified" India, the rebellion faded into the past. And without a direct military foe to face and rally against, the very purpose of colonialism was called into question. A justification was needed to explain why a diminutive island half a world away was controlling a totally foreign subcontinent, and that justification, as profoundly flawed and racist as it was, began to take the form of what Rudyard Kipling would refer to as "The White Man's Burden." The *true* purpose of colonialism never wavered—to exploit a foreign land's resources in the interest of the metropole. Yet in explaining it, British sentiment by the late nineteenth and early twentieth centuries was changing its tone, from that of pure conquest to one of protective custody—they envisioned themselves as harbingers of "enlightenment" and "civilization" to peoples that were, in Kipling's own words, "Half-devil and half-child." Almost as if the Europeans were doing their colonies *a favor* by subjugating them and putting them in their care.

This shift in colonial attitudes carried over to the realm of tiger hunting as well. Whereas the act of killing a tiger had once been almost synonymous with the military conquest of India, it became viewed in increasingly protective, paternalistic tones. For the white *sahib* by the turn of the century, killing tigers was considered a way of shielding the local Indian population from predation. It was, in effect, considered a benevolent act, a means of keeping rural villagers, who were generally depicted as "poor" and "helpless," safe from savage man-eaters. The "White Father," in all his supposed power and wisdom, was expected to defend local towns and villages from harm, to symbolically save India from itself, as it were. The presumption was preposterous—after all, rural Indian populations had been negotiating their existence alongside tigers long before the arrival of any Europeans. However, there may have been a kernel of practical truth, if for only the aforementioned reason: In the wake of the Indian Rebellion of 1857, most segments of Indian society were prohibited from having guns or weapons of any kind, without a difficult-to-obtain permit. Colonial officers of the early 1800s may have observed Indian villages banding together to drive away the occasional man-eater with spears, nets, and poisoned arrows, but such a thing would have been impossible by the advent of the twentieth century. Weapons had long since been outlawed, and much of the shared cultural knowledge of how to defeat man-eaters had been lost as well. If rural Indian populations had become helpless in the face of apex predators, it was largely because colonial policy had rendered them as such.

This attitude of British paternalism was most clearly depicted in the illustrations and hunting narratives of the era. As early as 1857, in a publication called *Tiger-shooting in India,* an Indian servant is depicted as cowering in terror before a charging, oversized tiger while a British hunter, on foot, no less, is shown gunning the animal down with calm determination. In a similar sepia print from a work published in 1871, titled *Wild Men and Wild Beasts,* Indian men are shown fleeing from an enraged tiger while a British hunter gets ready to slay the animal from an absurdly close range. Worth noting, of course, is that in both prints, the fleeing Indians are unarmed, while the British portrayed have guns at the ready—which is quite possibly the only part of the illustrations that's historically accurate. While the rural Indian population had generally been forced to surrender whatever meager arms they possessed following the Rebellion of 1857, the firepower available to British officials increased significantly in the decades that followed it. Prior to the 1870s, most European hunters in India relied on smoothbore, muzzle-loading muskets, which even sportsmen of the era decried as "very light, inferior [and] inefficient" when compared to the high-velocity rifles that appeared shortly thereafter. By the early twentieth century, many big-game hunters in India had adopted large-caliber double-barrel rifles like the 0.475-caliber H.V. This powerful gun weighed twelve pounds and could fire a one-ounce round at extremely high speeds, resulting in exceptional "stopping power." Other hunters preferred slightly smaller rifles, such as the 0.375-caliber Magnum, which fired a lighter bullet at a higher velocity—this was especially favored by trophy hunters, as it did less damage to animal hide. In either case, the accuracy, range, and power of the weapons available to colonial hunters improved dramatically. But even if Indians had been able to readily acquire the permits to own such weapons, they would have been well beyond the budget for all but the wealthiest members of Indian society. In 1909, a W. J. Jeffery 0.475-caliber high-velocity hunting rifle sold for around thirty-five pounds in London—this at a time when many indentured Indian servants were making the equivalent of around one pound per month. Even a significantly smaller sporting rifle, like a 0.256-caliber Mannlicher, would have cost the same indentured servant an entire year's salary. Not surprisingly, the average Indian villager or subsistence farmer was much more concerned with the immediate task of putting food on the table than the far more abstract and unlikely notion of shooting a marauding tiger. Which left dealing with such things almost solely in the hands of the privileged class—more specifically, the white *sahib*—whose "duty" to defend the unarmed rural poor gave him license to kill tigers as he saw fit.

Corbett surely would have been infected by these same ideas—as a member of colonial society he would not have been immune to the prevailing sentiments of the day, and even in his later writings about the Champawat Tiger, as we shall see, one can readily detect some of those same paternalistic undertones. But his sense of duty undoubtedly would have stemmed from genuine compassion for and fellowship with Indians as well. From his boyhood days in the forests of Kaladhungi, to his early years in the railway camps of Bihar, to his current posting at a ferry terminal on the Ganges, Corbett lived among the rural Indian population and considered them his friends and colleagues; he interacted with them and respected them in a way that was exceptionally rare among British colonists at that time. Indeed, as one of the few Europeans born in India, he considered himself *as* Indian, to whatever extent the ideologies of the day would allow, and as ambiguous as his own identity must have sometimes felt, he would not have hesitated to declare Kumaon to be his home. He had not, at least by that point, ever been to Ireland or England—India was all he knew. To Jim Corbett, the Champawat wasn't preying upon the foreign people of a distant land. It was killing and eating his fellow Kumaonis. Perhaps he did feel compelled as an aspiring *sahib* to perform his paternalistic duty. But he would have been equally motivated, perhaps even more so, by the very real fact that people he cared about—people "among whom I have lived and whom I love," by his own account—were being devoured in scores just a short distance from his hometown. After all, Corbett's life had been anything but easy: He'd given up on his dreams of becoming an engineer and dropped out of school as a teenager to help provide for his family; he'd spent more than a decade on his own at isolated frontier outposts, where men dropped like flies from cholera and malaria, and where human intimacy was an abstract notion at best; he'd lived through ample poverty, heartache, and suffering. No, not to the same extent that some of his Indian friends and colleagues had, but certainly more than any Europeans among his cohorts. The decision to go into the forest after a man-eater may have instilled in him considerable fear, but when asked, he didn't hesitate at the thought of sacrificing his own life for others. After all, it's what he'd been doing for almost fifteen years.

And if Corbett had indeed been waiting for the chance to prove himself, after his meeting with Berthoud, he wouldn't have to wait much longer. Just one week later, the word came, brought into town during the blackest hours of night by an exhausted runner on the verge of collapse. Between gagging breaths and choking heaves, the messenger would have struggled to spit out the news: the tiger had struck again. A woman had just been killed in the village of Pali, some sixty miles away.*

Corbett leapt into action. No minutes to ponder the decision he'd made, or second-guess the wisdom of going out to face such a creature. Man-eaters seldom stayed near a kill site for more than a few days, a week at most, and time was of the essence. He packed his things that same morning and set out for Pali, moving fast, traveling light, marching on foot through the sinuous mountain passes alongside six other Kumaoni men he'd recruited for the journey— not one of whom had any idea of the horrors that awaited them on the other side of those hills.

---

* It's worth noting that the village of Pali still exists, although it appears on maps as Pati Town. It has grown considerably since the days of Jim Corbett, with its concrete teahouses and food stalls making it a frequent rest stop for truckers. The original houses of the village can be found, however, just a short drive off of the main Almora road. Unlike the low-impact, temporary mud and thatch houses that the Tharu built in the *terai,* the Pahari people of the Kumaoni hills used loose stone to construct small huts or larger cottages, with roofs covered in irregular slate tiles. Many of these houses still stand today.

CHAPTER 6

DARKNESS FALLS

Jim Corbett and his men covered seventeen miles the first day alone, on their way to the village of Pali—up and down steep mountain trails, with packs laden with gear. Seventeen miles toward an animal that was believed to have devoured 435 souls. The march must have been tortuous, as they half jogged, half walked, drenched in the sweat of both anxiety and exertion. The hill folk they passed along the packed-dirt roads would have looked upon them with a mixture of suspicion and curiosity—suspicion as to why this gangly young railway worker with a mustache and short pants was marching into their mountains, curious as to the Martini-Henry rifle that protruded unavoidably from his pack. And these hill folk, these sari-clad women walking miles to find firewood, these topi-capped goatherds leading their famished flocks down from barren pastures, would have had their reasons. The year of 1907 was a tense one for Kumaonis, with their resentment over recent British forest regulations slowly simmering toward a boil. The Pahari hill tribes, much like the Tharu in the lowland *terai,* relied heavily on the forests for fodder, food, and fuel. The British, in attempting to preserve their crucial timber interests in Kumaon, had essentially declared most wooded areas as protected forests, and off-limits to all but British loggers and sportsmen. Organized forms of protest were beginning to take shape in the Kumaoni hills, including a massive demonstration in Almora that same year, to oppose the latest round of forest department regulations. Corbett does not mention such things in his account of the ordeal, but he must have been aware of the tension and the resentment around him. Tension and resentment that the untimely arrival of an almost supernaturally gifted man-eater would have only exacerbated. An angry forest goddess, perhaps, come to finally collect her due.

Panting, aching, Corbett and his men would have arrived in the village of Dhari that evening, setting up camp in the twilight to catch a few hours of sleep. If there was a government bungalow available, they may have spent the night there, although it seems more likely they slept in the open, beneath the stars. They weren't close enough to the tiger's hunting grounds to worry for their lives just yet, but they knew that was coming. The tiger's killing was still theoretical to them; its acts of predation were dry phrases, simple stories, devoid of shredded muscle tissue, bare of blood and bone. The reality of what the animal had accomplished in the hills of eastern Kumaon was still tinged with the sureality of myth—but that too was about to change.

After a quick breakfast at the nearby settlement of Mornaula, Corbett and his men set off again, resuming their furious pace through the mountains, covering almost *thirty* miles in a single day. Hillsides hatched with terraced fields, tawny green stands of serried pine, the white caps of the true Himalayas looming on the horizon—they would have seen all of this on the occasions when they looked up from the rutted path before them. The village of Dabidhura was approaching—the last stop before the killing site of Pali—and with its growing proximity surely came a mounting sense of foreboding. Although all present, including Corbett himself, were Kumaoni born and bred, one can easily imagine the disquieting feeling of marching into enemy territory, with every bush potentially concealing their foe, every wooded ravine the site of a possible ambush. In direct defiance of the latest British regulations, many hillsmen continued to burn out parts of the forest, both as a form of protest and as a means of regenerating wild grasses. In that dry month of May, it's not difficult to envision the almost apocalyptic image of smoke rising in black columns all around Corbett and his team, narrow mountain passes walled on either side by flame. Marching into darkness, into the smell of burning.

Another night of camping in Dabidhura, curried lentils cooked over an open fire, silent drags from hand-rolled cigarettes, and then they're off again, rising at dawn to cover the ten miles that remain between their campsite and Pali—the last stretch. Corbett says almost nothing in his writings about this final approach—perhaps there wasn't much to say. One can imagine the uncomfortable silence shared among the seven men, the pensive anxiety that consumed them as they climbed. Corbett, for the first time, may even have considered taking his rifle out of his pack.

And then they arrived. On the afternoon of May 3, 1907, with the sunlight showing the first signs of fading, they entered the village of Pali. Only, they felt as though they had entered a ghost town. No one greeted them. The slate-roofed stone huts were all silent; the central courtyard bare. Local custom would normally dictate a welcome from the headman, and the ceremonial offering of sweets and hot tea, but no such reception materialized. And just the arrival of a European—an oddity many deep in the hills had never seen—was usually novel enough to summon at least a band of curious children. But no, they didn't appear either. In fact, nobody appeared, and it became evident, almost immediately, that something terrible had happened here; that this was a haunted place. They looked around uneasily, muscles tensed, unsure of what was happening. And there came a smell as well—a gag-inducing stench, not of rotting flesh, but of excrement—the cause of which became clear as soon as Corbett and his men dropped their packs and built a fire.

What occurred next would stay with Corbett for years, and indelibly shape his understanding of man-eaters. It was to be his first encounter with actual victims, and his first clear picture of the psychological trauma that their communities incurred. One by one, the villagers emerged from their darkened stone houses, "in the state of abject terror," as he would later write, ghost-eyed and visibly shaken. Although reluctant at first, the gathered residents of Pali, shivering in their soiled clothing, slowly began telling the new arrivals what had happened to their village. This is what Jim Corbett would remember, when writing down his recollection of events some years later:

> I was informed that for five days no one had gone beyond their own doorsteps—the insanitary condition of the courtyard testified to the truth of this statement—that food was running short, and that the people would starve if the tiger was not killed or driven away. That the tiger was still in the vicinity was apparent. For three nights it had been heard calling on the road, distant a hundred yards from the houses, and that very day it had been seen on the cultivated land at the lower end of the village.

The arrival of Corbett's hunting party seemed to have reassured the fifty-odd residents of Pali, at least enough for them to come out from behind locked doors—something they had been too afraid to do for almost a week. However, when Corbett asked to be taken to the site of the last kill, they were understandably reluctant. The tiger was still out there, and the woman it had taken was hardly its first victim in the area. The villagers were all too aware of what this creature was capable of, and that Corbett was not the first shikari sent in to stop it. With the sun setting and darkness closing in, taking this strange Kumaoni-speaking Englishman out to see the tiger's latest feeding grounds was out of the question. They did, however, eventually give Corbett a detailed account of how the most recent victim had been killed, and it was a story all too familiar in the realm of man-eating tigers.

A group of women, some twenty in number, had been out at the edge of the forest, collecting oak leaves to feed to the village cattle. One of the women had decided to climb a tree to harvest extra leaves, and she was in the process of climbing back down when the tiger attacked, rearing up on its hind legs and ripping her out of the tree, with a violence that left the other women stunned. Before they had time to even react, the tiger had switched its grip to her throat and gone barreling away, back up the steep side of a ravine and into the undergrowth. The terror-stricken women sprinted back to the village to get help, and a group of men did band together to go after the tiger, perhaps even attempt to save the victim, although the hopelessness of that task became quickly apparent. The tiger had already begun feeding in a dense cluster of laurel, and upon their approach, the creature charged, unleashing an eardrum-shattering roar that scattered the unarmed party and sent them fleeing back to their homes. A century before, a village like Pali might have possessed the weapons and the martial know-how to stop such an animal—spears, nets, poisoned arrows—but by 1907, a full fifty years after the armed uprisings of 1857, such weapons had long since been taken from them by the colonial government. The lone resident of Pali who had even some semblance of a gun had fired the old blunderbuss in the air when the tiger appeared, more confident in its abilities as a noisemaker than a lethal weapon. Needless to say, the strategy had not worked well. The tiger had not abandoned its kill, and the search party had not been able to stop it from feeding.

Huddled around the fire with the assembled villagers, Corbett listened closely to their stories and faced the uncomfortable decision of what to do next. As a lifelong hunter, dangerous predators were hardly new to him; he'd had hair-raising run-ins with bears, leopards, and tigers before he was in his teens. Not being a trophy hunter, however, he generally did what any sensible Kumaoni would and avoided them altogether. But now, for the first time in his life, he found himself in a role to which he was totally unaccustomed. He was seeking to actually confront not only a feeding tiger, but a man-eater at that. This *was* a novel endeavor, and there was certainly no instruction manual or guidebook upon which he could rely. By his own admission, "there was no one I could ask for advice, for this was the first man-eater that had ever been known in Kumaon; and yet something would have to be done." Corbett wasn't exactly right on this count—the rare man-eater does crop up in the historical records of the region. But there certainly had never been a serial man-eater quite like this one. Corbett was no stranger to your average tiger—he had, in fact, access to a body of firsthand, indigenous knowledge that would take Western naturalists decades to catch up with—but now he faced an extraordinary creature. He would have to learn the delicate art of hunting a man-eater on the fly, and it was a subject with a steep and unforgiving learning curve. Mistakes were all but inevitable; he could only hope they would not prove lethal in the end.

This lack of experience, coupled with the natural folly of youth, might explain why Jim Corbett chose, that same night, to commit what was quite possibly the most foolhardy act in his entire hunting career. Once he was assured that his six companions were safely indoors for the evening, he decided to spend the night *outdoors,* on the ground, on the chance that he might get a shot at his target. Wild tigers have a peculiar penchant for man-made roadways, and Corbett's suspicion that this tiger might as well was confirmed by the people of Pali, who told him that the man-eater had been seen stalking the road outside the village at night. Given that it was a full moon, and visibility would be good, Corbett eschewed the relative safety of a stone shelter for a single tree by the side of the dirt thoroughfare. Rifle in hand, back to its bark, he hunkered down for the night and waited—wondering if he would even have time to get off a shot should the tiger decide to make him its next meal. It had been at least five days since its last kill, and Corbett was well aware of the tiger's need for a weekly feeding. The time to hunt again was dangerously close.

As the moon rose and the chill gathered, it became increasingly clear to Corbett just what a mistake this strategy had been. Although aggressive man-eaters have no qualms killing by day, tigers are by nature nocturnal hunters. With retinas made up primarily of light-sensitive rod receptors, and a reflective tapetum lucidum layer right behind them, the tiger has night vision some six times more acute than that of a human. And coupled with those powerful eyes is an equally impressive pair of ears. Built on a swivel, they can readily hone in on a source like radar dishes, and have an impressive range of between .2 kHz and 65 kHz—considerably broader than the range for humans, whose extends only up to 20 kHz—geared toward picking up noises as faint as the swallowing of saliva and the whistling of breath through nostrils. In addition to vision and hearing worthy of a superhero, tigers also have five different types of whiskers covering their body, all of which are constantly absorbing sensory information and empowering them to navigate through the darkest and densest of underbrush. When one considers that this surveillance package comes mounted atop an animal that is virtually silent thanks to padded feet, and camouflaged to blend in perfectly with the banded and stippled shadows of night, one can't help but realize—just as Corbett did soon after taking his post—that trying to outmaneuver a tiger once the sun has set is as suicidal as it is naïve. In his blithe determination to set up a trap, it had not occurred to him that he was not the deadly snare so much as the helpless bait.

But he didn't go back. He couldn't. His strategy may have been foolish, but it was all he had, and Corbett knew that running back to the village would have meant losing the faith of its inhabitants. He needed to prove to them, and possibly even to himself, that he could do this. Corbett did come from a military family, after all. Both his father and grandfather had served in a host of wars and campaigns, and as a boy, Jim had been raised on tales of sacrifice and valor. No, turning tail was out of the question. He had made his choice. As to the terror that followed his decision not to run, who better but Jim Corbett himself to describe it:

> The length of road immediately in front of me was brilliantly lit by the moon, but to the right and left the overhanging trees cast dark shadow, and when the night wind agitated the branches and the shadow moved, I saw a dozen tigers advancing on me,

and bitterly regretted the impulse that had induced me to place myself at the man-eater's mercy. I was too frightened to carry out my self-imposed task, and with teeth chattering, as much from fear as from cold, I sat out the long night. As the grey dawn was lighting up the snowy range which I was facing, I rested my head on my drawn-up knees, and it was in this position my men an hour later found me—fast asleep; of the tiger I had neither heard nor seen anything.

When Corbett returned to the village for breakfast, bleary eyed, the night's chill still clinging to him, the local men were surprised he had made it out alive—as was he, with the benefit of hindsight. It seems likely that the tiger, still very much in the area, would have been aware of his presence. Yet in explaining its reluctance to attack, one need only consider the motivations of a tiger. Even a man-eater will usually only kill if it feels threatened or is sufficiently hungry. In his crouched, motionless position against the tree, Corbett would have posed little obvious threat, and it seems—fortunately for Corbett—that the tiger was still adequately sated from its last victim to forgo hunting for the night. But the tiger was almost certainly near, and it very well may have been watching him from the edge of the forest, those golden eyes turned to silver in the moonlight, on the brink of a pounce, muscles coiled, crouching but then hesitating, suddenly uncertain—then finally turning away and melting back into the shadows.

ııııııııııııııııııııııııııı

Over the next day, as more and more villagers shared their stories with Corbett, the true gravity—and enormity—of the situation became apparent. The residents of Pali were still too afraid to tend

to their fields or gather feed for their animals. They were too terrified even to walk the roadways to seek help from their neighbors.
They had become veritable refugees in their own homes, stalked
by a specter that seemed able to kill them at will. And far from
being a local phenomenon, this experience was common to villages
across eastern Kumaon. The entire countryside was paralyzed, as
no one knew where or when the tiger might strike. In rural communities where simply relieving oneself involved a walk into the
woods, the inhabitants couldn't help but feel that they were constantly at the mercy of the tiger. And all of this in a mountainous
region where tigers were hardly known at all. Leopards and bears,
yes, they did prowl the rugged highland terrain, but tigers were
creatures of the jungled lowlands. For the hill-dwelling Kumaonis,
a man-eating tiger was a new kind of terror altogether.

This sense of foreignness has persisted, I learned, even into the
modern day. While most of the lowland-dwelling Tharu I spoke
with while researching this book had heard tigers before and could
easily reproduce their roars and calls, the Pahari people I met in the
hills of Kumaon were at a total loss when it came to tigers—they
could imitate the grunts of leopards and growls of bears with expert
precision, but the sounds of the tiger were completely unknown
to them. It was then, as it is today, a creature alien to their world.

However, Corbett's show of bravery that night, foolish as it was,
did appear to have reaped some benefits. The headman of the village asked him if he might be willing to stand watch over the
village wheat fields so the crops could be harvested and the animals
grazed; it seemed the farmers, who had virtually no rifles or guns
of their own, had been encouraged enough by Corbett's display
of bravado to make a tentative return to their work, providing he
was there with his rifle in hand. That at the very least meant there

would be something to eat, in a village that had gone days without food. Recognizing the seriousness of the situation, Corbett immediately agreed to the request. After executing a preemptive search around the village for any fresh pugmarks, which turned up a heart-stopping eruption of kalij pheasants from the bushes, but thankfully no tigers, he set up watch beneath a walnut tree as the people of Pali returned to their work. By evening, the crops from five large fields had been cut with sickles and bundled into sheaves.

Whatever confidence daylight had given them, however, was quickly rescinded with the coming of the dark. Following a quick cleanup of the courtyard, the people of Pali locked themselves in their homes and huddled around their cooking fires. The foul odors of waste had at the very least been replaced by the familiar smells of Pahari cuisine—there was wheat again for chapati, and milk to make ghee—but the fear remained: Corbett was still unable to convince anyone in the village to take him into the forest where the last victim had been killed. And it was a fear that had begun to infect him as well, although possibly for the better. He had survived one brazen night in the open by the skin of his teeth, and he made the wise decision not to press his luck. His second night in the village, he stayed indoors, to sleep beside a blazing hearth fire. The door of his hut was packed with thick thorn branches to keep the man-eater from entering, and a loaded rifle was at the ready in case it should try. Somewhere outside the tiger lurked, prowling, waiting, its hunger mounting, its next hunt coming closer by the hour.

||||||||||||||||||||||||||||||

Up until Corbett's arrival in Pali, the notion of a man-eater had been abstract. The tigers he had known before that moment behaved by a certain set of rules; a sort of biological etiquette whose contours followed the general shape of the natural world. The tigers Corbett was familiar with would occasionally snarl or engage in a bluff charge to scare off a human who had stumbled upon a fresh kill, but they didn't crush a person's neck between their jaws and strip the flesh from their limbs. The tigers of his youth might skate the edge of a village to snatch a goat, but they didn't march up into the mountains and lay siege to a town like an invading army. No, this thing he was chasing seemed a different animal entirely, one whose natural instincts had become so perverted and twisted as to render it in some ways unrecognizable as a tiger at all.

As to whether Corbett understood, as early as 1907, the central role human beings had played in its perversion is difficult to say—although he must have realized that something unprecedented was taking place. In fact, the arrival of a new class of man-eater in the region was even beginning to gain traction in major colonial newspapers of the time, well beyond the hills of Kumaon. In an editorial published that same year, *The Times of India* would lament how such predators had "infested the jungles all this time around Dhunaghat, Devidora, Lohaghat, and Champawat," with the troubling addendum that "hitherto the authorities have entirely failed to do anything for the protection of the villagers." And it is a claim backed up by the government records of animal attacks at that time. Of the forty-two human beings killed by tigers in the United Provinces in 1907—a year that was not even half over when Corbett's ordeal began—thirty-nine occurred in the Kumaon division alone. Most of the other divisions recorded no fatal tiger attacks at all, and Meerut, Agra, and Fyzabad claimed just one victim each. Clearly, what was happening in Kumaon with its thirty-nine victims was a bloodbath in comparison. Essentially, it seems that a single tiger—the very tiger Corbett had been commissioned to destroy—was responsible for 93 percent of the fatal tiger attacks in an Indian territory larger than the state of Wyoming. And keep in mind, the actual number of its victims was almost certainly much higher, due to a lack of official reporting in remote villages, and to the fact that local British officials would not have been eager to showcase just how inept their campaign against a recalcitrant Indian tiger had been.

Corbett understood that without the trust of the residents of Pali, people who "knew every foot of the ground for miles round," his mission would fail. He urgently needed to see the tiger's tracks—and the only place Corbett knew he would find them was in the forest, where the tiger had taken the woman. For an expert tracker like Corbett, a single set of pugmarks from a tiger could provide a wealth of information, a sort of natural résumé that detailed the critical facts needed to plan the hunt. From the shape of the toe pads, he would be able to determine if the cat was male or female—those of males tend to be larger and more circular compared to those of females, which are generally more delicate and elongated. He would also be able to determine the tiger's age by how developed or worn the toe pads were, detect potential injuries by the tiger's gait, see whether or not there were cubs present, and collect at least some basic information on its feeding habits. And there was far more information to be gleaned beyond the animal's tracks; claw marks, urine patches on trees, and fresh droppings could all yield precisely the sort of intelligence that Corbett needed to formulate a strategy. One of Corbett's unique talents, both by his own account and the testimony of others, was an uncanny ability to imitate the calls of wild animals. This made for a nice parlor trick in Nainital, but in the jungle, it was an indispensable tool for attracting game. By knowing more about the tiger, he could determine how best to find it, to stalk it, and potentially call it out into the open, simply by re-creating the mew of a lost cub, or the grunts of a female in heat, if appropriate.

All of which was contingent upon being led to the site where the tiger had taken its victim—something the residents of Pali still steadfastly refused to do. The true depths of their terror is perhaps best revealed by the fact that they were devout Hindus, and recovering whatever human remains they could find for ceremonial cremation was absolutely essential; that the woman's friends and relatives were still too shaken to venture into the forest to perform a religious duty speaks volumes on the psychological damage the tiger had caused. The people of Pali were deeply traumatized, as one would expect of a population that had lived for days with the striped specter of death hovering constantly over them. They had seen someone they loved ripped from a tree and dragged screaming into the forest; they had spent sleepless nights listening helplessly to the murderer's roars as it circled their village, knowing it had been feeding on their departed friend. They were shaken to their core, and without some form of reassurance, nothing would convince them to seek out a second confrontation. Corbett was at an impasse; crucial minutes were ticking by, and soon, the tiger would either kill again in Pali or simply move on to hunt somewhere else.

In his account of the hunt, Corbett divulges little about the backgrounds of his companions. But he must have trusted them— he did choose them for a quest to kill a man-eater, after all—and it stands to reason that he would have valued their opinion. They may not have been rural hillsmen like the residents of Pali, but as Indians, they had experience and insights that a European like

Corbett, despite his grasp of the Kumaoni language and culture, simply did not. Nainital, where they came from, was essentially a segregated town, and they were no strangers to the ways colonialism could engender fear and misgiving. It was likely their own appreciation for the villagers' doubts regarding colonial authority—coupled, perhaps, with the grumbling in their own bellies—that gave birth to a plan. Something that might put some confidence in the people of Pali, *and* some victuals in the pot, with little more than a few well-placed shots. A show of marksmanship, as it were. Proof that this foolhardy Britisher, sent by a government they had no faith in, actually had the skills to protect them. Corbett approached the headman of the village shortly after, rifle in hand, and asked where best to find *ghooral*—the spry little mountain goats that Kumaonis across the board considered something of a delicacy. The headman apparently approved of the idea, and upon announcing the plan to find fresh game, three brave locals volunteered to take him there. The *ghooral* could be found on the steep grassy slopes just outside the village, and the men were understandably excited, after a week without food, to finally have something substantial for their families to eat.

As the people of Pali watched from behind half-closed doors, Corbett and his three new guides marched boldly together to the outskirts of the village. They crossed the main road and made their way down a steep ridge, until they reached a point some half a mile away where a series of ravines converged. Surrounded by gulches that could have easily concealed a tiger, their attention was drawn to the sudden appearance, high up on a distant hill, of a gamboling *ghooral,* its head poking out from a patch of wild grass. It was a tremendously difficult shot—Corbett estimated the mountain goat to be close to two hundred yards away, with just its head showing, at an awkward sixty-degree angle. But he knew it would be the only chance he got.

Although it was a boast he would never admit to in writing, Corbett was an exceptional marksman. He had honed his aim early on, while pursuing game in the jungles of Kaladhungi. When his family fell on hard times following the death of his father, bush meat became an important part of their diet, and head shots were crucial, as not to spoil an animal's flesh. His marksmanship had also earned him praise and respect as a young student—he joined Nainital's Voluntary Rifles regiment as a cadet at the age of ten, and after impressing a visiting sergeant major with his rifle skills, was allowed to borrow a .450 Martini carbine for his own expeditions into the forest. The breach-loading Martini-Henry rifle was a weapon he would quickly master, and it was the same weapon, among others, that he would bring with him on his mission some twenty years later. And although somewhat heavy and prone to recoil, Corbett would admit that it "atoned for its vicious kick by being dead accurate—up to any range."

As for *ghooral,* however, that remained to be seen. Lying down flat against the ground and balancing the barrel on a pine root, Corbett took the shot. The hilltops resounded with the clamor of gunpowder, and for a moment, there wasn't anything to see but smoke. The three men from the village strained their eyes to catch any sign of a wounded animal, but there was nothing—they murmured among themselves in disappointment, certain that Corbett had missed.

But he had not. As Corbett reloaded, the limp form of the *ghooral* came sliding from the tall grass, then began to tumble down the steep bank of the hill. In its fall, however, it came to disturb two other *ghooral* goats from a second patch of scrub, and as they sprang into view, Corbett found himself aiming once again, getting a bead and squeezing the trigger, praying that he could accomplish "the seemingly impossible," as he deemed the feat he was hoping to perform. Two more bursts of black powder, and somehow, two more limp and tumbling *ghoorals,* one shot through the back, the other through the shoulder. The bodies of the three *ghoorals* somersaulted down the full length of the hill before coming to rest at the bottom of the ravine, directly in front of Corbett and the villagers. For an instant, man-eating tigers were forgotten and they cheered at the meat that had landed before them. They quickly gathered up their trio of *ghoorals* and made their way back to the village, where the people of Pali had assembled in the main courtyard, eager to greet them.

As the *ghoorals* were skinned and dressed, Corbett overheard his guides telling the other villagers wildly exaggerating versions of what had occurred: tales of magic bullets that could kill an animal from a mile away, and summon it to land before a hunter's feet. Normally, Corbett would have dispelled such talk with a wave of his hand, but on this occasion, he wisely chose to let these "*shikar* yarns," as he deemed them, percolate throughout the village—to fortify their courage, and perhaps even his own. They all were going to need it.

Following a communal midday lunch of curried *ghooral,* Corbett asked once again to be taken to the tiger's feeding site, and this time the people of Pali, emboldened by Corbett's sharpshooting and no doubt grateful for the first substantial meal they'd had in some time, obliged. The headman helped assemble a group of volunteers, and from them, Corbett selected two of the men who had accompanied him on the *ghooral* hunt as guides—a charge they now enthusiastically accepted. As they made their way through the village, toward the waiting forest beyond, the family of the woman who had been taken approached, asking them to bring back any remains they might find so that they could be cremated in accordance with Hindu custom. It was a request that Jim Corbett humbly agreed to, and he promised he would do his best.

The three men left whatever protection the tight ring of stone houses offered, walking slowly past the terraced fields and entering tentatively into the first line of trees. They must have been conscious of the distinct disadvantage they were placing themselves at as they were enveloped by the forest.

Normal, wild tigers will generally surrender a kill site when confronted by humans. With a snarl, perhaps, even flattened ears and a display of teeth—but they seldom persist beyond that. Maneaters like this one, however, had no such compunctions. Like the ultra-aggressive cats that would appear in Chitwan National Park a century later, this tiger had virtually no fear of humans and would yield nothing; a procession of hunters was little more than a cavalcade of meat. And if it chose to strike, there would be only fractions of a second to react to the blinding assault. A single shot, if Corbett were lucky, and while his aim was exceptional, it often took more than one bullet to stop an enraged tiger. This was why most tiger hunting in India involved shooting from the safety of a howdah atop an elephant, or from a machan stand high up in a tree—staying beyond the reach of their claws and teeth was crucial. When conducted on foot, tiger hunting was a very different occupation, with a margin of error so slim as to be virtually nonexistent. At the very least, a dog trained for tiger hunts could afford some protection, both by detecting tigers early by their smell, and by serving as a convenient distraction should a tiger attack. But as a newcomer to the vocation of tracking man-eaters, Jim Corbett had none of these advantages. There was no *hattisar* near to lend him an elephant, no time to build a machan and set out bait, and no hunting dogs in the vicinity to sniff out the trail. To walk right into this tiger's feeding site was to court death in its most disturbing form.

Still, the trio of men possessed their own advantages. For starters, it was early May, which meant that the summer monsoon, although just around the corner, had not yet arrived to choke up the valleys with impenetrable foliage. It was because of the monsoon that tiger hunting in northern India was generally a winter occurrence—in the warm, wet months, the grasses and thickets became so dense as to render all tigers virtually invisible, and impossible to find, let alone shoot. In colder, dryer weather, the forests lost much of their vines and creepers, creating a window in which hunting was feasible. Second, Jim Corbett had an extensive body of knowledge in regards to tigers, and while this particular tiger broke with many conventions, there were some that he was certain it would still abide by. Corbett knew the feeding habits of tigers, and he understood that once a large animal was killed, the tiger would spend several days in a row intermittently feeding and resting, seldom venturing far from the carcass, at least during the daylight hours, until all the edible portions were gone. Granted, almost a full week had passed, and it was increasingly unlikely that the tiger would still be nearby—but if it had not yet left on another hunt, it would almost certainly be close to the original feeding site, either quietly finishing its meal or drowsing away the afternoon in a shady spot just beside it. This knowledge may have provided Corbett with some comfort, but probably not much. Unlike his companions from Pali, who had never seen a modern rifle fired before that day, he knew there was nothing magic in the bullets he carried; if the tiger became aware of them before they saw it, there was a very good chance he wouldn't have time to get off a shot. Just a flash of stripes before the claws did their work. One more addition—or perhaps even three—to the tiger's unconscionable tally.

All of this would have cycled through Corbett's mind as he ventured closer to where the tiger had fed. Before visiting the site, however, the hunter asked to be shown where the woman had been attacked, hoping to reconstruct in his mind exactly what had occurred. He needed to learn as much as he could about this particular tiger—how it hunted, how it killed. His companions led him to the oak tree in question, where he saw at last the first tangible evidence of an attack.

A patch of dried blood marked where the tiger had sunk in its teeth and dragged the woman from her perch—with such violence and force, evidently, that shreds of skin still clung to the bark from the palms of her hands, a sight that seems to have shaken Corbett greatly. Struggling to keep his composure, he retraced the tiger's approach from a nearby ravine, and found, at last, a pristine set of pugmarks in the fine earth that had settled between two rocks. Finally, he knew something about the tiger—not much, but a clue nonetheless. It was a female, good-sized, and seemingly in good health, although a little past its prime. Given that wild tigers can live up to fifteen years, and in some cases even longer, this seemed to corroborate what Corbett had suspected about its history as a man-eater. If the tiger had been killing humans for roughly eight or nine years in both Nepal and India, and if it had been a young adult when first wounded in the lowland *terai,* then its age would have likely been in the vicinity of ten to twelve years—making it an

older tiger, although still more than capable as a hunter. And the fact
that it was a female could also help to explain its initial transition
to man-eating; the pressures of feeding cubs early in life, especially
if natural prey was inaccessible, could have easily led a mother tiger
to consider slow-footed and weak-limbed Homo sapiens as food.

A trail of dried blood led from the oak tree back down into the
ravine, where the tiger had apparently taken its victim, and this
Corbett and his two guides followed. Eyes wide, ears straining,
they descended into the shadows, the occasional russet spattering
of blood their only guideposts. If the safety catch on Corbett's rifle
was not released before, it certainly was by the time they reached
the bottom. But the tiger did not stop here; it had taken the woman
even farther, up the opposite bank, to a dense cluster of bushes
where it had finally stopped to feed. The three men approached
the final site in silence, with aching slowness, Corbett's rifle surely
raised, its barrel used cautiously to part the curtains of leaves and
peer within . . .

Blood was everywhere. But no tigress. According to Corbett,
the animal had eaten virtually all of the body; nothing remained
after days of feeding but a few shards of bone and shredded clothing. With as much reverence as they could muster, the three of
them assembled whatever meager remains they could find and
wrapped them in the clean cloth that the victim's family had given
them to serve as a shroud. Then, confident but not yet certain that
the tiger had moved on, they re-created their own somber version
of a funeral march—the men cradling what was left of their beloved neighbor for cremation, Corbett holding his rifle just in case
her killer should return—back to Pali, where they faced the heart-
breaking task of presenting what could barely be called a body to
her grieving family. But it was enough for cremation, for the sacred

rite to be performed, and for the ashes of the tiger's victim to reach "Mother Ganges."

|||||||||||||||||||||||||||||||||

If Corbett's first two days in Pali had been a crash course in man-eaters, the next three proved to be a study in frustration. Having yet to confront the tiger, he was still acting as something of a detective on the trail of a serial killer. Back in the lush *terai* forests near Kaladhungi, he could have relied on the testimony of the "jungle folk," as he liked to call them, to help pinpoint a suspect tiger. The belling of sambar, the barking of *kakar* deer, and the hooting of langur monkeys were all clear indicators that a tiger was near; the calls of birds such as scimitar babblers and blue magpies could readily provide information on where a predator had passed. And if all else failed, there was always the congregations of blowflies and vultures that seemed to follow in a tiger's bloody wake. But here at altitude in the Himalayan middle hills, the lessons he had picked up while tracking in the lowland jungles were of considerably less value—this was not ideal habitat for a Bengal tiger. There seemed to be little beyond tangled scrub and chill ravines, places where the only sound was of one's own heartbeat, and of the unsettling winds that sighed through the pines.

Corbett did experience some hope of a lead, however, upon discovering a living witness residing just outside the village who had seen the tiger up close. A young girl and her older sister had been attacked while cutting grass the year before, with the elder of the two being carried off into the forest. But the experience had so traumatized the surviving sister—she had actually chased the tiger with a sickle until it turned to roar back at her—that she was either unable or unwilling to talk about it. From the rest of the family

Corbett was able to gather few details about the animal that he did not already know. A pattern, however, presented itself: women and children were making up a disproportionately large percentage of its victims—an unsettling realization for anyone, let alone a man who had essentially been raised by his widowed mother and older sisters. It did make a cruel sort of sense, though, given the division of labor common in the Kumaoni hills at that time. While men toiled in the fields and older women prepared meals and cared for animals, it was usually the responsibility of the young women and children to go into the surrounding forest—often in direct defiance of government forestry regulations—for the crucial tasks of gathering firewood and animal fodder. Which meant, inevitably, that they bore the brunt of the tiger's attacks.

Corbett spent from sunrise to sunset for three straight days wandering the forests that encircled the village, senses on high alert, rifle at the ready, asking everyone he met if they had seen or heard the tiger. He was pointed to watering holes where the tiger was thought to have drank, and the shady bowers where it was believed to have rested, but each new lead proved to be a dead end. No fresh pugmarks, no scat, no tiger. It seemed as if the creature had vanished into the ether just as inexplicably and unsettlingly as it had materialized before.

One name, however, kept cropping up. In the whispering of the villagers who beckoned to him from their windowsills, in the stories of the farmers who peeked out from behind closed doors, one location was mentioned repeatedly—a place where the tiger seemed to call home, and where, according to their tales, it had committed its most brazen acts of killing. Here was where the creature always returned—a place so dangerous, simply walking alone in the open was considered an act of madness.

Champawat. A larger town some fifteen miles away, the nexus around which all of the tiger's hunts seemed to hang in dark orbit. It was the closest Corbett had to a viable lead, and the one trail that he knew he could actually follow. After three fruitless days of searching, he had no other choice. It was clear the tiger was no longer near Pali. There were no more tracks that pocked the dust of the road, no more roars from the forest that ripped out the seams of the night. The tiger was gone, simple as that.

Corbett gathered together his men from Nainital and made the announcement—they would pack up their things and leave for Champawat at dawn.

CHAPTER 7

||||||||||||||||||||

TOGETHER, IN THE OLD WAY

The jungles of northern India were hardly a place for the inexperienced or the uninitiated. Poisonous snakes, fractious sloth bears, lurking crocodiles, aggressive elephants—not to mention protective mother tigers—all could and occasionally did prove fatal to humans. And as a boy growing up in Kumaon, Jim Corbett had mentors who taught him the art of survival in the forest. His cousin, Stephen Dease, an amateur naturalist, gave Jim his first gun, a derelict muzzle-loader, in exchange for helping him collect specimens of local birds. There was Dansay, the disinherited son of a general and fellow Hibernian who entertained the young Corbett around the campfire with tales of Irish banshees and Kumaoni *churails,* forever mingling in his mind the superstitions and folklore of his two homelands. And of course there was Jim's older brother Tom, whom he worshipped like a hero and did his best to imitate on their earliest hunts together for peafowl.

Of all Jim Corbett's teachers, however, none would equal the lasting and profound influence of Kunwar Singh. In addition to serving as the headman of the village of Chandni Chauk, just a short hike from the Corbett family's winter lodgings near the *terai,* Kunwar was also, in Jim Corbett's own words, "the most successful poacher in Kaladhungi." As to what he meant by the term "poacher" isn't clear—game restrictions were still few and far between at the time—although it seems Kunwar didn't have much regard for the Indian Forest Act of 1878, and even less compunction about hunting in the protected timber reserves that the colonial government had deemed off-limits. His defiance of such policies could have been simple economic necessity, but there is also a revealing bit of information to which Corbett makes only a passing reference. Kunwar Singh was of Thakur descent—essentially the Kumaoni equivalent of minor nobility—and it's very likely that prior to the arrival of the British, those same "protected" forests he was trespassing upon had served as his family's hunting grounds for generations. To such a man, hunting for food and sport among the same sal and haldu trees as his ancestors would not have been "poaching" at all, but rather the exercise of an ancient birthright—one he refused to yield to the foreign rulers of an alien government. A Kumaoni Robin Hood, you might even say.

With all this in mind, a friendship between a Thakur elder and an eight-year-old colonial boy must have seemed highly unlikely. Yet Kunwar not only took a young Jim Corbett under his wing, he actually became a father figure after the boy's true father passed away. Such a relationship would have been next to impossible in the provincial capital of Nainital, where a state of social segregation between Europeans and Indians was actively enforced. However, in the scattered villages around Kaladhungi, where the Corbetts spent the non-malarial winter months, such strictures held little sway, and the family mingled freely with their Tharu and Pahari neighbors, Hindu, Muslim, and Christian alike. It was "Uncle" Kunwar who came to check in on a young Jim Corbett after the boy narrowly escaped a run-in with a leopard, and who subsequently began taking his inexperienced friend along with him on hunts, to impart as much of his knowledge of the natural world as he could. And upon the reception of Jim's first gun—an important rite of passage in the local culture—it was Kunwar who would tell him: "You are no longer a boy, but a man; and with this good gun, you can go anywhere you like in our jungles and never be afraid."

Under Kunwar's tutelage, Jim learned to see the forest through the eyes of an indigenous shikari rather than those of a European hunter. The plants and animals had a language all their own, and Kunwar Singh instructed the young Corbett in its lexicon and grammar. Jim soon was hunting like a true Kumaoni: stalking barefoot through the Garuppu jungle, avoiding foxes, which could curse a hunt, reading the flights of birds to track recent kills, burning out *nal* grass to reveal hidden game, and staying in the forest's good graces with blessings of the peepal tree. When it came to the ways of the jungle, Kunwar Singh was a teacher without equal, and Jim Corbett a more than eager student.

There was one lesson, though, one story, that stuck with the boy—that had instilled caution in him during his earliest hunts, and that surely did so again, years later, as he marched along that dark and uncertain road to Champawat. It was a warning of sorts, a tale from Kunwar's own experience that he had imparted on Jim to forever remind him of just how quickly fortunes could turn in the forest. Before telling him the story, however, Kunwar had given the young man the same advice still heeded by many forest dwellers in tiger-prone areas of India and Nepal today: "When in the jungles, never speak of a tiger by its name, for if you do, the tiger is sure to appear." With that warning, he began his story:

In the month of April of the previous year, Kunwar had gone into the jungle with his friend Har Singh, to hunt food for their respective tables. Kunwar was by far the more experienced of the two, having spent significantly greater time tracking game in the surrounding forests. But Kunwar was happy to have a companion, and the two men stalked quietly through the tall grasses and dense foliage of the Garuppu jungle, ever-mindful of the colonial forest guards who patrolled the region for poachers, and always on the lookout for the armed bandits, or dacoits, who were known to use the woodlands as a hideout from authorities. There were a host of potentially dangerous animals as well, although Kunwar had far less fear of them than he did of man—he knew their habits, and when to steer clear.

The first ominous sign arrived in the form of a jungle fox that crossed their path as they were leaving the village. Kunwar, with all the experience and wisdom of a Kumaoni shikari, immediately recognized the bad omen and suggested they turn back—it only signified trouble in the offing. His friend Har merely laughed at his old-time superstitions, saying it was "child's talk" to think that a harmless little fox could ruin a hunt. To a town dweller like Har, such a thing was patently absurd. Against his better judgment, Kunwar gave in, and the two men continued their march deeper into the dense scrub and thorn bamboo of the jungle.

Kunwar missed his first shot at a chital stag feeding in the pale morning light, and Har Singh lost a wounded peafowl in the grass not long after. Both should have been easy shots, but something had gone unaccountably wrong—it seemed as if their bullets had been cursed from the onset. After an utterly fruitless day, the two men ultimately decided it best to head back; nothing else appeared in terms of game, and they were both concerned that the shots they had fired earlier may have alerted the forest guards of their presence. With the afternoon turning quickly to evening, they followed the course of a nullah, or dry creek bed, to avoid the trails that the forest guards usually patrolled—with their illegal, unlicensed rifles, denying that they were poaching would have been next to impossible.

They were right to be cautious—their gunshots had been noticed, although not by any human. No, in this case, they had summoned a tiger. Its massive form emerged from the leaves and stood staring at them, golden eyes aglow in the mossy twilight. The men froze; they dared not move. For a terror-filled minute they simply stood still, locked in its gaze. Then, as abruptly as it appeared, the tiger turned tail and vanished into the forest. Speechless and somewhat shaken, the men continued on their way, at a pace that was no doubt considerably accelerated—night was coming, darkness was on its way. And there had been something in the tiger's posture that had left Kunwar deeply unsettled. He had run into tigers plenty of times before, but this one felt different. Its peculiar stare felt almost like a warning, a spine-tingling omen of a danger to come. Perhaps there was a fresh kill nearby, or maybe it had cubs—he didn't know, and he didn't want to find out. The message it was communicating, however, was undeniably clear. *Turn around. Go home. Today, this is not your forest to hunt.*

The two men continued on their way, walking uneasily along the nullah rather than taking the road, and sure enough, the tiger appeared once again, materializing from the dense foliage that abutted the creek's sandy bank. Only this time, it was visibly agitated: its striped tail twitched in obvious displeasure, its flanks rumbled with the profound beginnings of a growl. Once again, the two men froze, and once again, the tiger glared at them in menace before vanishing back into the dark shadows of the leaves.

At which point it became clear to Kunwar that the natural

balance of the forest was horribly askew. Between the fox, and their missed shots, and the sudden appearance of an irritated tiger, he knew they needed to get back on the road and out of the trees as soon as possible. The spirits were out of sorts. On that particular day, the two men did not belong there.

At that very moment, however, a flock of jungle fowl rose before them, with one alighting upon the branch of a haldu tree only a few feet away. It was an easy shot—free food for the taking— and after a day of such dismal luck, Har Singh simply could not resist. Kunwar tried to stop him when he saw the rifle come up, but his warning came a second too late. The sharp report of the rifle was answered directly by an unimaginable roar, and just as he had feared, the now furious tiger came crashing in toward them through the brushwood. Kunwar, with his ample jungle experience, knew exactly what to do. Unlike leopards, tigers are relatively poor at climbing trees, and they will seldom pursue a human who has surmounted the reach of their claws. At the first sight of its barreling stripes, Kunwar scrambled up the nearest *runi* tree, its forked trunk and rough bark leaving plenty of purchase for his bare and callused feet. Har Singh, on the other hand, was neither so well versed in the ways of the jungle, nor so lucky. The comparative city slicker was still scrambling for a branch to hold on to across from Kunwar when the tiger sprang at him. It was not a predatory attack—simply a defensive one, of the sort that involves roars and claws rather than spine-severing bites to the nape of the neck. But still, this was more than enough. Kunwar watched in horror as the tiger reared upon its hind legs, pinned his companion to the trunk of the *runi* tree just opposite his, and amid a mingled tumult of snarls and screams, proceeded to eviscerate the poor man with its claws.

Kunwar had his rifle with him up in the tree and did consider taking a shot at the tiger, but he quickly realized he risked shooting Har Singh. If he did nothing, on the other hand, he knew that his friend would soon be dead anyway. So Kunwar did what he thought best and fired his rifle into the air. Fortune, for once in the day, appeared to be on their side; this time, at the sudden sound of a gunshot, the tiger fled, and Har Singh collapsed in a bloody heap.

Kunwar waited a silent minute to be sure the tiger was gone before descending from his perch and approaching his friend, who was shuddering and moaning at the base of the *runi* tree. Upon turning him over, Kunwar discovered just how ferocious the attack had been. In addition to shredding most of the bark and outer wood from the trunk of the tree, one of the tiger's claws had also entered Har Singh's stomach, tearing the lining "from near his navel to within a few fingers' breadth of the backbone." In just a few short seconds, the tiger had gutted the man like a fish, and his intestines were spilling out onto the jungle's leafy floor.

Remarkably, Har Singh was still conscious, and in fierce whispers the two men debated if the jumbled mass of innards should simply be cut away or pushed back inside. Har Singh felt strongly that his own guts should be put back where they belonged, and Kunwar, although certainly no doctor, was inclined to agree. Working in silence in case the tiger was near, Kunwar stuffed the mass of intestines back into his friend, including the "dry leaves and grass and bits of sticks that were sticking to it," before wrapping his own turban securely around Har Singh's middle to hold it all in.

The worst of it may have been over, but their ordeal was far from finished. Night had fallen, and the nearest hospital was still ten miles away, which meant an excruciating and terrifying hike was ahead. Despite his catastrophic wounds, Har Singh walked the entire distance back in the darkness with Kunwar leading the way, the latter shouldering both of their rifles in case the tiger returned to finish the job. The hospital was closed when they finally arrived, but luckily the doctor was still awake, and he attended as best he could to the injured man. With Kunwar holding the flaps of his friend's stomach together, and a local tobacco seller steadying a lantern for whatever light it could give, the doctor stitched up the hole, twigs and all, with nothing more than a glass of liquor for Har Singh to help kill the pain.

Ironically, it was not Har Singh who would meet a premature death. Despite his grievous injury and slapdash medical treatment, the man made a full recovery, living to a ripe old age and passing away many years later of natural causes. It was Kunwar Singh— Corbett's old friend and mentor—who expired far sooner than he should have. The expert shikari and once-proud Thakur, the headman of his village, fell victim to the same cycle of substance abuse and addiction that plagued so many indigenous communities displaced and dispossessed by the forces of colonialism. With his privileges gone, his ancestral hunting grounds off-limits, his stature diminished, the shikari sought comfort in the cheap opium that flowed in from China and Tibet. The last time Corbett had seen his old friend, he was emaciated and strung out, lying at death's door on the filthy mud floor of a servant's shack. Corbett tried to nurse his "Uncle" Kunwar back to health, and even had him swear an oath, upon a sacred thread and a leaf of the holy peepal tree, to give up the powerful narcotic for good. And Kunwar apparently kept his oath—he did not die that day. But he would never hunt again as he once did, and his damaged body finally gave out a few years later. The community lost a village elder, one of the last true shikaris with a native knowledge of tigers, and Jim Corbett lost a father for a second time.

It must have occurred to Corbett, as he raced down the narrow dirt paths through the hills toward Champawat, that if his friend were still with him, they could smoke tobacco once again around a campfire and plan the hunt together. Kunwar could offer him wisdom and advice: Think like the tiger, decipher its language and its intentions, the way a true shikari of Kumaon must. But Jim Corbett was on his own. And even if his old hunting companion had been present to offer him guidance, Corbett surely suspected what his counsel would have been: *Turn around. Go home. This forest and this tiger are not yours to hunt.* Respecting nature and not tempting fate were the core lessons of the story Kunwar had told him as a boy, and it was the sort of wisdom that allowed a wise shikari to live another day.

But Corbett also knew that another day meant another victim. Another woman out cutting grass, another man tending to his fields, another child gathering wood for a fire. Another blood trail that led to a shadowy ravine full of scattered gore and splinters of bone.

Darkness was closing in, and the many scars on the pine trunks where the Kumaonis had tapped them for torch resin would have testified to just how profound that darkness could be. Jagged black hills, lightless ravines, the empty roar of rushing water—night was falling, just as it had for Kunwar when he met his tiger back in the nullah. Only rather than fleeing from it as any smart shikari would, Edward James Corbett was hurtling toward it.

|||||||||||||||||||||||||||||||||

Much about the man-eater may have been unknown to Jim Corbett, as he marched with his small party into the tiger's favored hunting grounds, but details of its existence—hints, perhaps, at even how it might be found—were beginning to reveal themselves.

He may not have had the expert advice of his friend and mentor Kunwar Singh, but he possessed the cumulative experience of an entire lifetime spent in the forests of Kumaon, and an almost preternatural insight into the habits of the animals of the region, a sort of "sixth sense," by his own account, that enabled him to see the natural world through their eyes and predict their movements. It would have occurred to Corbett, after just the few days he had spent on the tiger's home turf, what made this animal such a proficient killer of men. It wasn't that this tiger was exceptionally bloodthirsty or murderous, or even vastly divergent in terms of feeding habits from its more conventional tiger kin. No, it was that this tiger had developed a hunting strategy that, when coupled with the region's geography, made it almost impossible to pinpoint. True, when it came to prey, the tiger had totally lost its fear of humans; but at the same time, it had learned to avoid them in the immediate aftermath of a kill. And in terms of its territory, now free from the rivalry of its fellow tigers, it had accomplished the rare feat of drastically expanding its holdings. Its evolved hunting methods seemed to resemble more closely those of an Amur tiger of the Russian Far East than a Royal Bengal—it patrolled a very large area, covering a huge swathe of eastern Kumaon, and it never seemed to stop moving for very long. Amur tigers employed this strategy due to a scarcity of prey—finding food in cold northern forests demanded an all but nomadic existence. In the case of the Champawat Tiger, however, a shortage of prey was not a concern. Rather, it seemed the Champawat adopted this strategy specifically to avoid the hunters who sought it. Perhaps it was a lesson learned from that first wound to its jaw, or possibly wisdom acquired after its close call across the border in Rupal. But either way, it had absorbed the lesson well. Essentially, the Champawat had become a hit-and-run artist, a tiger that snatched villagers from the edge of the forest, devoured them quickly, and then moved on. Kill sites could easily be twenty miles apart, and there appeared to be no way to predict exactly where it would strike next—beyond the fact, of course, that the tiger seemed inexplicably compelled to return periodically to the environs of Champawat.

And that, ultimately, was what accounted for its tremendous human tally. It did not kill faster or more effectively than other tigers, nor wantonly, for that matter—its weekly hunts were more or less on par with what any wild tiger would accomplish. But rather, this tiger was almost impossible to find, let alone stop, and it had continued killing, unhindered, for a long stretch of time. The remoteness of the region, coupled with the fact that most locals were prevented from owning firearms, certainly didn't help. Even an immediate response following an attack on a human meant sending a runner to the nearest large town with a colonial government presence, an endeavor that could easily take several days. Then, an experienced hunter needed to be found and dispatched, which could take several more days beyond that. And by the time that hunter arrived—often a week later or more—the tiger would have already finished feeding and moved on, ready to attack anywhere in a vast territory of steep ravines and rocky hills. In the past, when villages had the weapons and knowledge to stop a tiger themselves, and when regional maharajas still had the authority and means to assist, it would have been a problem solved at the local level. Given the centralized nature of the colonial government, however, hunting such a tiger in such an inaccessible region had become an exceptionally difficult affair.

This behavior on the part of the tiger not only meant that a would-be hunter was always one step behind—which accounts for the failures of the various paid shikaris, forestry officials, and soldiers sent in previously to stop it—it also compelled the local population to live in a constant state of terror. Tigers certainly don't have the capacity to commit acts of psychological warfare, but that was precisely the effect that the tiger had upon the people of Kumaon. And as Jim Corbett neared Champawat, this fact was demonstrated in a new and unsettling way.

As Corbett passed the village of Dhunaghat on his way to Champawat, he began encountering large groups of people traveling on the road. At first, he may have assumed they were pilgrims on their way to a shrine, or possibly relatives journeying together for a festival. But the tone was off, and there were no shrines in the vicinity, nor was it the season of festivals. When his curiosity could bear it no longer, he approached one such group of men, some twenty strong, and asked them why they were traveling together in such large packs. The answer should not have come as a surprise, but it did—*the tiger*. The roads in the area were considered so dangerous, locals only felt safe traveling them by daylight, and only in groups of a dozen or more. Simply visiting neighbors, or going to a bazaar in a nearby village, had become a communal, almost martial affair. People marched in roving platoons, always wary, their eyes alert for striped fur through the trees, their ears attuned for distant roars.

As Corbett soon learned, just two months earlier, some of the men in this very party had witnessed firsthand what the tiger could do. While on their way to a market in Champawat, they'd heard an agonized cry from a valley below. The screams continued to mount and grow closer, until the tiger emerged from the tree line with a woman, still alive and pleading for help, held in its jaws, its teeth locked around the small of her back. From a mere fifty yards away, they watched as it crossed the road before vanishing once again into the forest.

Confounded, Corbett asked if they had attempted to stop the tiger. Indeed, they had—the men from the group sprinted to the nearest village to gather reinforcements, including several poachers who had unlicensed guns. The assembled rescue party, now fifty or sixty in number, followed the blood trail along the length of the valley, banging drums and firing muzzle-loaders in the air as they did so, hoping to scare the tiger away from its victim—which they actually were able to accomplish. But for the young woman, who had been gathering firewood when the tiger attacked her, it was too late. She was gone—she had been stripped of her bloody clothes by the tiger, and robbed of her life. Ashamed and humbled, averting their eyes from her body, the men used their own unwound dhoti cloths to preserve whatever remained of her dignity, and carried her back to her relatives in the village.

Corbett garnered two crucial insights from encountering this group of travelers. First and foremost, that he needed reinforcements just for the journey; even the road was not safe from the man-eater. Upon revealing his purpose—to hunt and kill the tiger in question—the local men, still disturbed and enraged by what they had seen, agreed to accompany him to Champawat and help in whatever way they could. Second, from talking to his new companions, it became all the more evident that stopping this tiger was not going to be a solitary endeavor either. Whatever illusions Corbett had harbored about besting the creature single-handedly were precisely that. Hunting it effectively would mean shedding the old Anglo-Saxon mythos, the need to confront Grendel alone in its lair, and instead think—and *hunt*—like a Kumaoni. It meant working as a team and a community, to finally bring almost a full decade of killing to an end. It meant hunting together, in the old way.

And so this unlikely party, this mounting army, now nearly thirty members strong, began marching in loose formation through the chilly dusk. Twenty rugged Pahari hillsmen, six porters from Nainital, and one khaki-clad Englishman, all banded together in a common purpose.

No, it was not Jim Corbett's tiger to hunt. It was *their* tiger to hunt. All of them, as Kumaonis.

The traditional tiger hunt, or *bagh shikar*, of the Indian nobility was conducted with bows and spears, and generally executed at a sustainable level. The hunt itself had ritualistic importance.

With the arrival of the British, the nature of the Indian tiger hunt changed. High-powered rifles replaced more traditional weaponry, and tigers were slaughtered wholesale from the backs of elephants. This photograph, taken in 1911, shows a colonial-style tiger hunt organized for King George V in the Nepalese *terai*.

Around 1907, reports began circulating throughout northern India of an exceptional man-eater: the so-called Champawat Tiger, responsible for more than four hundred deaths. The image of the tiger as a blood-thirsty predator was widely propagated during the colonial era, with their alleged aggressiveness used as an excuse to justify their eradication. The Champawat, however, was one of the few tigers that actually lived up to that reputation, as an injury sustained by a poacher's bullet combined with a severely degraded habitat had forced it to begin hunting humans to survive.

A rare photo of Jim Corbett as a young man, taken while serving as a humble railroad employee, roughly around the time he first heard mention of the Champawat Tiger. The Corbett family was well-respected in Kumaon, by the Indian and British communities alike, although their Irish origins and limited finances kept them out of the upper echelons of the colonial elite. Jim was considered "country bottled," a derogatory term used for European settlers born in India. *(Courtesy of the Jim Corbett Museum in Choti Haldwani)*

The natural habitat of the Royal Bengal tiger is the *terai,* the belt of marshy jungle that lies at the foot of the Himalayas. The Champawat Tiger almost certainly began its life in the *terai,* a place rich in game and mates. *(Dane Huckelbridge)*

Author's photo of a pugmark in Nepal's Chitwan National Park. Tigers are elusive predators, with the occasional footprint or dropping being the only hint at their presence. *(Dane Huckelbridge)*

The *terai* has been occupied for millennia by the Tharu, an indigenous ethnic group that has always held the tiger in high esteem. It was traditionally the role of the *gurau,* or village holy man, to execute the proper puja sacrifices and keep the forces of nature in harmony. If the spirits of the forest were offended, it was believed that a tiger could be an instrument of their revenge.

This photograph, taken during an expedition to Kumaon just prior to Jim Corbett's hunt for the Champawat Tiger, shows the stark contrast between the porters from Nainital and the topi-capped hill dwellers, as the true Himalayas tower in the background. Corbett's hunting party was composed of both groups.

The cold, pine-capped foothills of the Himalayas are nothing at all like the lush jungles of the *terai,* but this was exactly where the Champawat Tiger would eventually hunt the bulk of its human victims. This photograph was taken by the author on the outskirts of Champawat, close to where the tiger killed its final victim. *(Dane Huckelbridge)*

This is one of the few surviving photographs of the Champawat Tiger's taxidermied head, and convincing evidence of why it began hunting humans in the first place. Clearly visible are its missing lower canine and damaged upper canine teeth, an injury Corbett attributed to a poacher's bullet early in life. *(Courtesy of the Dalmia Family, Gurney House)*

The bottom of the gorge where Corbett squared off and killed the Champawat Tiger. The author (middle) was taken there by two expert guides, both of whom knew local families that had participated in the hunt back in 1907. *(Dane Huckelbridge)*

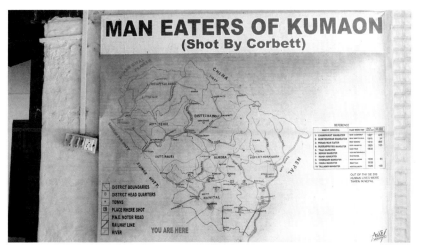

MAN EATERS OF KUMAON
(Shot By Corbett)

After bringing down the Champawat Tiger, Jim Corbett acquired a reputation as the leading hunter of man-eaters. This ability served him well, at a time when deforestation and diminishing prey were driving more and more tigers and leopards to hunt humans as food. *(Courtesy of the Jim Corbett Museum in Choti Haldwani)*

Although occasionally misidentified as the Champawat, the tiger pictured here beside Corbett is the Bachelor of Powalgarh, shot in 1930, the largest cat Corbett would ever hunt.

Jim Corbett devoted the latter part of his life to tiger conservation. As an early proponent of protected tiger reserves, he played a crucial role in saving the species from extinction. This photograph, taken in the national park that today bears his name, shows a majestic specimen of the animal that Corbett loved above all others: the Bengal tiger. *(SunnyMalhotra/ Shutterstock)*

CHAPTER 8

IIIIIIIIIIIIIIIIIIIIII

ON HOSTILE GROUND

On February 7, 1815—the same year that the East India Company took Kumaon from the Nepalese Gorkhas, who in turn had taken it from the Doti kingdom that preceded them—newlyweds Joseph and Harriet Corbett arrived in the British colony of India, following a grueling six-month voyage from their home in the British colony of Ireland. They hailed from Belfast, a city at the center of the political and religious violence that wracked the north of the troubled island. Although Joseph Corbett would list "gilder and carver" as his trade, it appears his true training was of a significantly *higher* order. According to at least one account, Joseph had previously been a monk—and Harriet, his wife, had formerly been a novice at a convent that was, evidently, temptingly close by. The couple left the monastic life, eloped, and in doing so, almost certainly put themselves in a situation that was scandalous at best, and flat-out dangerous at worst. In the north of Ireland at that time, disavowal of the Church would have been seen as a provocation, for Catholics and Protestants alike—a dangerous proposition in a place where colonial hatreds were ancestral, and rebellions and uprisings a grim fact of life. Just in his own lifetime, Joseph would have witnessed the Irish Rebellion of 1798, Michael Dwyer's guerrilla campaign of 1799, the Irish Rebellion of 1803, and the Castle Hill Rebellion of 1804. In short, the Ireland he knew was an unforgiving place, one where the breaking of religious ties and sectarian allegiances could prove to be a perilous transgression. And in rebuking the Church, that seems to be precisely what the erstwhile monk Joseph and the nun-in-training Harriet had done. No longer welcome among the Irish, and not ever having been trusted by the presiding English, the young couple apparently decided their only chance was to seek their fortune several oceans away, on the other side of the globe.

If they felt any relief upon escaping their situation back in the Old Country, it was to be short-lived. In attempting to get out of their rock-and-a-hard-place jam in Ireland, they had, metaphorically speaking, jumped out of the frying pan and into the fire. Malaria, cholera, and dangerous animals aside, it quickly became clear to the pair that they had come upon a land that was just as cloven by sectarian divides as the one they had left, and equally unforgiving of those deemed on the wrong side. Indeed, Joseph's only escape route out of Ireland had been to sign up for unlimited service as an infantry private. The British had a long history of recruiting their colonial subjects to serve in the far-flung outposts of their empire, and in what to Joseph must have felt like a particularly cruel form of irony, he quickly learned that he had fled the frequent insurrections and rebellions of one British colony, only to end up risking his life subduing them in another. With the war with Nepal still in full swing upon his arrival—the same conflict that would eventually win for the British control of Kumaon from the Shahs—young Joseph was thrust into almost immediate service in India's volatile northwestern frontier. Regardless of his prior standing or stature, however, opportunities did exist in India that were simply unavailable to an uneducated Irishman back on the old sod. Joseph seems to have done relatively well for himself during his time in the military, being promoted to the rank of sergeant in the horse artillery, building a modest home in the outpost of Meerut, and fathering nine children with Harriet before meeting his own premature demise at the age of thirty-three, possibly from malaria acquired while out campaigning, although one can't be sure.

The sixth of those nine children, Christopher William, grew up to follow in his father's martial footsteps—likely because there were few other realistic options available for a domiciled colonist of Irish descent. For those of English birth and aristocratic lineages, opportunities in the civil service and officers' corps abounded; for a fatherless young man of lowly rank, however, enlisting was perhaps the best way to ensure a roof over one's head and three square meals. He was able to climb the ladder slightly, however, and was chosen to be trained as a junior medical officer, achieving by the age of twenty the rank of assistant apothecary with the 3rd Troop of the 1st Brigade, of the very same horse artillery with which his father had once served. He would see action in the First Afghan War of 1839, and even earn a medal for distinguished service in the Kabul campaign. The work was not highly skilled—much of it, in fact, would have consisted of holding down wounded patients and hacking off limbs—but it was a step up from the sort of "cannon fodder" roles on the colonial frontier that many young men of his background were destined to fill. Christopher William Corbett would go on to serve with distinction in the East India Company's various expansionist endeavors, including the Sikh wars, for which he was awarded the Sutlej Medal in the Aliwal campaign in 1846, and later in the Punjab campaigns, where he served as a hospital steward in the Bengal Army. He was briefly married and fathered three children, although his wife, like so many others unsuited to the foreign diseases and rigors of colonial life, passed away in her early twenties. A heartbroken Christopher William soldiered on.

Of all the horrific campaigns and battles Christopher William would witness, one in particular would have a lasting impact on his own life, and those of colonists in India as a whole. To the British, it was known simply as "The Mutiny"; to Indians, it would eventually be heralded as the First War of Independence. To all parties involved, it was an exceptionally violent conflict, marked by bloody insurrections and even bloodier reprisals. What began in 1857 as a relatively minor disagreement regarding gun grease— local Hindu and Muslim sepoy troops employed by the East India Company were offended by rumors that the lubricant contained beef and pork fat—would ignite a hidden powder keg of colonial tensions, culminating in an open insurgency. Some groups, such as the Sikhs and Pathans, remained loyal to the British. Others— particularly those aligned with dynasties such as the Mughals that had suffered the most under British rule—sought violent retribution. And while rebellions broke out in scattered locations across the whole of India, in the northern frontier the fighting was especially fierce, and atrocities were committed wholesale by both sides. On several occasions, the "mutinous" sepoys shocked British sensibilities by massacring the colonial populations of entire towns—the Siege of Cawnpore and the Bibighar Massacre were two of the most horrendous examples—and the British troops responded in kind with their own campaign of slaughter and terror. Villages were ransacked, entire towns were burned, and suspects were killed indiscriminately. The true scale of the carnage is difficult to describe, let alone imagine, although the following eyewitness account of the British attack on the rebel town of Jhansi at least hints at the horror:

> Fires were blazing everywhere, and although it was night I could see far enough. In the lanes and streets people were crying pitifully, hugging the corpses of their dear ones. Others were wandering, searching for food while the cattle were running mad with thirst . . . How cruel and ruthless were these white soldiers, I thought; they were killing people for crimes they had not committed . . . .
>
> Not only did the English soldiers kill those who happened to come in their way, but they broke into houses and hunted out people hidden in barns, rafters and obscure, dark corners. They explored the inmost recesses of temples and filled them with dead bodies of priests and worshippers. They took the greatest toll in the weavers' locality, where they killed some women also. At the sight of white soldiers some people tried to hide in haystacks, in the courtyards, but the pitiless demons did not leave them alone there. They set the haystacks on fire and hundreds were burned alive . . . .

The spasms of violence that the rebellion set off would affect colonial families all across India, and the Corbett clan was no exception. Christopher William served throughout the conflict, and managed to survive it, although others close to him were not so lucky. His younger brother, Thomas Bartholomew, was captured in Delhi during the Siege of Red Fort, tied to a tree near its gates, and burned alive for all to see. And the first husband of the woman whom Christopher William would later go on to marry—Mary

Jane Doyle—was killed at the Battle of Harchandpore. According to surviving accounts, he was pulled from his saddle and hacked to pieces in combat, while Mary Jane and her children were suffering through their own trials miles away, trapped and on the verge of starvation in the Siege of Agra.

The Rebellion of 1857 was eventually put down, although India and the Corbetts both would be forever changed. In the smoldering aftermath of the uprising, the British Crown took it upon itself to impose full, martial control over India, and Christopher William took it upon himself to ask the freshly widowed Mary Jane Doyle, who had survived the siege at Agra, to be his wife. She said yes, and he decided he'd had quite enough of fighting for the British. In 1858, the newly married Christopher William retired from army life and took up working for the postal service instead. The pay was minimal—particularly given that he now had a total of six children from previous marriages to support—but being hard up for money surely sounded better than being hacked to pieces or burned at the stake. He worked as a postmaster in the towns of Mussoorie and Mathura, until a transfer in 1862 finally took him to the picturesque hill town of Nainital. With its shimmering lake and pine-capped peaks, it was a lovely place to raise a family—although at its high elevation in the foothills of the Himalayas, also unbearably cold in the winter. Upon the advice of District Commissioner Sir Henry Ramsay—the same Sir Henry Ramsay who would initiate many of the controversial forestry regulations of the 1860s and '70s—Christopher William saved his rupees and built a small Irish-style stone cottage at Kaladhungi, on the edge of the *terai* at the base of the hills, where the family could stay during the cooler, non-malarial months of the year. With a little extra income coming in from real estate ventures—Mary Jane had a knack for such things—and a reputation in the community as something of a town leader, Christopher William surely took pride in what he had accomplished, and some comfort in the fact that the Corbett family was finally established.

However, the halcyon days were not to last. Regrettably for the Corbetts, the family's fortunes would once again take a downturn. Christopher William, who had survived battles and skirmishes and rebellions galore, would be killed at a relatively young age instead by a degenerative heart condition that had gone undiagnosed for years.* But not before fathering *nine* more children—the youngest of whom, almost thirty years after his father's death, would find himself walking with Martini-Henry rifle in hand into what was not only the hunting grounds of the deadliest tiger in history, but also one of the last true hotbeds of anti-colonial sentiment in India. Memories of the rebellion persisted in the Corbett household—Jim's older brother Tom had been named after Thomas Bartholomew, the uncle who had been burned alive by rebels in Delhi—and Jim would have known that Champawat had been one of the epicenters of conflict in the United Provinces. During the uprising, Kalu Mahara, one of the rebellion's principal leaders in the northwest, had rallied the people of Champawat and its environs to his cause, promising independence for Kumaon

---

* The death of Christopher William was the low point in a traumatic period for the Corbetts. A massive landslide in Nainital the year before had also killed many of the family's friends and neighbors, and prompted the Corbetts to relocate their lodgings to Gurney House, a cottage on the other side of town. The death of Christopher William marked the beginning of a sort of genteel poverty for the clan, with money being tight for years to come. The family was able to make ends meet, thanks to Mary Jane's real estate work and some additional postal income from Jim's older brother Tom, but the Corbetts' economic and social position was severely affected by their father's untimely demise. Jim's childhood, though generally normal and happy by his own account, would be forever marked by the tragedy, and by the economic hardships that ensued.

after almost seventy years of combined Nepalese and British dom-
ination. Capitalizing on decades of simmering resentment, Kalu
Mahara organized a militia from among the Pahari hill tribes and
began leading guerrilla-style attacks against the British barracks
in Champawat's vicinity, including fortifications at Lohaghat, and
even the much larger station at Almora. Caught off guard by the
violence, colonists living in the area fled alongside a tide of other
refugees to the relative safety of Nainital, farther west. The British
troops, though initially scattered by the attacks, were eventually
able to regroup, and with the help of reinforcements from neighboring towns managed to stage a series of effective counterattacks.
In fact, it was Commissioner Ramsay, the old acquaintance of the
Corbett family, who was in charge of organizing the forces to put
down the rebellion. In the end, Kalu Mahara's campaign failed,
and he was arrested along with many of his fellow "mutineers."
Accounts vary as to what happened next—some say he was executed, others claim an angry mob freed him from his prison cell in
Almora. Either way, his cause had been crushed, and with it, the
dream of an independent Kumaon. But if memories were long in
the Corbett household, they were even longer in Champawat—a
place where thousand-year-old temples were actively venerated,
and ancient battles still celebrated in song—and Kalu Mahara's
sacrifice was not forgotten. Anti-British sentiment, though common throughout India, ran especially high in the region.

Of course there was more to that animosity than the fate of
one folk hero, and one can easily find its more immediate sources
in the draconian laws that were passed in the rebellion's wake. At
least some of the people's resentment would have stemmed directly
from the government's general prohibition of firearms—a state of
affairs that had been officially in effect since the Indian Arms Act

of 1878, passed after the rebellion to severely regulate the possession of weapons among the Indian population. The inevitable result, particularly in more volatile frontier districts like Champawat, was the clandestine possession of illegal guns. Granted, the sort of technologically advanced, high-velocity rifles available to British colonists were difficult to come by, so most contraband weapons were outdated and of poor quality—the sort of weathered muzzle-loaders that Jim Corbett himself had first hunted with as a boy. Some of these weapons were indeed used for the illicit purposes the Crown feared most; the vast majority of unlicensed guns, however, would have served far more pedestrian purposes, akin to those of Corbett's old friend and mentor Kunwar Singh. Namely, the pro-curement of wild game from the forests, protection against dacoits, and as insurance for livestock and kin against the rare incursion of a problematic predator. Not that such weapons were generally very effective against a five-hundred-pound tiger charging at forty miles per hour, but they were better than nothing—which was still what the vast majority of the population had on hand. Guns were, regardless of their legality, considered a highly valued asset in Champawat's frontier culture.

With the revolts of 1857 still vivid in the collective British imagination, however, the idea of an armed Indian populace was a frightening prospect, even half a century after the fact. While Europeans were exempt from such measures, acquiring a legal firearm, even for defense of livestock or protection against wild animals, was difficult under the terms of existing legislation. In 1905, an article in *The Times of India* recorded a total of 8,901 new licenses for firearms, issued in accordance with the Indian Arms Act of 1878. This number may sound substantial, but one must consider that the total population of India at the time was verging on 300 million—effectively, the percentage of the population with legal access to firearms was minuscule, intended by design to be virtually nonexistent. And from another article in *The Times of India,* dated April 15, 1907, it becomes clear why that number was kept purposefully so low:

> Diligent students of newspapers in this part of the world can hardly fail to have been struck by the fact that fire-arms are now being frequently used in the commission of crime, says the "Englishman." They have been produced in the case of riots, and within a few days no less than three cases have been reported of persons shot dead by others who ordinarily should not have been in the possession of rifles or guns. When a Maharaja, particularly a friend of Europeans and officials, is shot from behind a hedge and the Police Superintendent of a District has a bullet whistling over his head, the time has come to enquire by what means criminals or fantastic persons on this side of India manage to possess themselves of fire-arms . . . All this points to the fact that a demand for weapons has suddenly arisen in Bengal. One would naturally like to know why. Some people will find no hesitation in accepting the reply that the demand has been caused by those Bengali newspapers and other preachers of sedition, who proclaim that the people of this country ought to perfect themselves in military exercises and the use of arms.

Striking, perhaps, is the author's incredulity and mild outrage at the fact that a people living beneath the yoke of a foreign power have the audacity to procure weapons for any reason, let alone self-defense. It is a view tinged with paranoia, but not without at least some basis in reality, as without a doubt, there was no shortage

of individuals in Bengal who had implicitly good reasons for re-
senting the British Empire, and even better ones for wanting to
defend themselves from it. Of course, Kumaon was not in Bengal, but concerns similar to those expressed in the article would
have no doubt been shared by the colonial government and population. And as Jim Corbett was soon to discover, contraband
guns—although generally of poor quality—were indeed owned in
considerable numbers, in Champawat in particular. It's difficult
to imagine, however, that Corbett, a lifelong resident of Kumoan
who was well versed, nay, *steeped* in both the history of the 1857
uprising and the illicit gun culture of the region, did not have at
least some idea of the type of place he was stepping into.

Of course, he says little on the subject in his account of the
hunt, which is not surprising—Corbett studiously avoided "pol-
itics" in his writing, and he often presented a considerably more
sanguine vision of colonial life than history can account for. Which
isn't to say he lied or stretched the truth—the man was known for
both his humility and his honesty, by Indians and Englishmen
alike. It's more a question of reading between the lines. Like so
many colonial narratives, both in the Old World and the New,
his accounts of living and hunting during the Raj do conveniently
elide over some of the uglier facets of the experience. A sort of selec-
tive memory, one might even say, where the horrors of a tiger attack
are vividly remembered, while the horrors of colonial violence are
conveniently (or perhaps even necessarily) forgotten. Regardless,
however, of what is included and what is omitted in his accounts,
one has to assume that Corbett was well aware of Champawat's
historical relationship with the government, and of the people's
deep mistrust of foreigners such as himself. Descending upon the
town with his companions in formation, he was a long way from

the manicured cricket greens and sunny tea parlors of Nainital—a town largely unscathed by the violence of fifty years before. In a place like Champawat, on the other hand, the rebellion might as well have been yesterday, and the sight of an armed "Englishman" leading loyal "native" troops into town would have been not only disarmingly familiar, but downright upsetting. The sort of glares Jim Corbett likely encountered on its outskirts may have been relatively novel for him, but they would have been all too familiar to his grandfather Joseph, from his days back in the bitter lanes and crumbling shanties of County Antrim. All across Kumaon, in both the hills and the *terai*, government forestry regulations and labor policies were causes for popular resentment; in Champawat, however, massacres and mass executions were still part of living memory. It very well may have dawned on Corbett that a man-eating tiger was not necessarily the most immediate threat to his safety. If northern India could be compared to Northern Ireland, Champawat was its West Belfast—and Jim Corbett, regardless of his intentions, was entering it as an armed emissary of the colonial British government.

As he crested the ridges on the outskirts of town, curling along the rims of the first terraced fields, edging along the ancient stones of the Baleshwar Temple, it must have occurred to Corbett that unlike those in Pali, it would take much more than a display of trick shooting at *ghoorals* to win over the people of Champawat. For unlike Pali, Champawat was no diminutive village—it had a population of several thousand. And the town itself was of no small historical significance. According to devout Hindus, Champawat was where the *Kurmavtar*, or the turtle incarnation of the god Vishnu, had originally manifested itself, and the surrounding valley had served as the seat of the once-mighty Chand dynasty

for more than five hundred years. Champawat's days as a religious
and municipal capital had long since passed by the time Corbett
marched through its figurative gates, although the town's richly
painted facades and elaborately carved balconies would have at-
tested to its former glory.*

Its residents were proud and defiant, if weary of living for the
better part of four years alongside a tiger that devoured their loved
ones on a regular basis. The curious arrival of Corbett and his
miniature army of followers must have surely brought them to
their windows and doors, perhaps even coaxed a few contraband
muzzle-loaders out from beneath floorboards or straw piles. Who
did this scrawny Englishman think he was, marching into *their*
town with a ragtag bunch of illiterate hill people, and six turbaned
dandies from Nainital?

It must have been a pressing and somewhat intimidating
question—no doubt posed amid the snarling of stray dogs and the
warning tolls of Baleshwar's bells. But fortunately for Jim Corbett, he had someone to vouch for him. He arrived with a letter
of introduction, almost certainly from the headman back in Pali,
who'd had the foresight to recognize the dubious reception he might
encounter in the far-eastern borderlands of Kumaon. The letter was
intended for the Tahsildar of Champawat, a man whose proper
name Corbett does not mention, although an article published
later in the Indian journal *The Pioneer* would identify him as
Pandit Sri Kishan Pant. One can assume he was a respected and

---

* Champawat, much like Pali, has grown significantly since the days of Jim Corbett, including the addition of modern concrete structures. However, on the preserved streets near the
Baleshwar Temple—which itself dates back to the twelfth century—one can still see vibrant
traces of the Old City, as Corbett would have seen it. With its narrow passages and ornate
balconies, Champawat's historic architecture is distinct from that found in the other towns
and villages in the area, and harkens back to its days as a seat of dynastic power.

educated individual of advanced years, likely hailing from a well-established, high-caste family. Rather than the common topi cap and coarse woolen cloak donned by most hillsmen, he probably wore an honorary turban, or pugaree, together with a long embroidered kurta robe to denote his special status. The title of Tahsildar in India dated back to the early days of the Mughal Empire, and it derived from both Arabic and Persian to essentially mean a collector of revenue. And while the role did indeed originally entail the collection of taxes for the king, it was also considered a position of leadership within the community—a more formal and officially recognized version of the village headman.

The Tahsildar was not present when Corbett first arrived in Champawat, and the hunter proceeded to do what almost any employee of the colonial government would have done, particularly one on his own in a historically hostile part of the country: he quickly repaired with his men to the dak bungalow. At the time, it was commonplace for any sizable Indian town to have a special structure set aside for visiting officials, a well-kept albeit somewhat Spartan guesthouse. The structures were a common feature of colonial life, although they had a darker, less obvious significance as well. During the Rebellion of 1857, many British colonials, fleeing from rebel-occupied towns and fortresses, had used these dak bungalows as emergency shelters, and more than a few had met a violent end within their walls—often by fire, which was why thatch of any sort was banned as a building material in such bungalows following the revolt. The bungalow's ominous connotations permeated the colonial consciousness, even revealing itself in the stories of Rudyard Kipling, who would write: "a fair portion of the tragedy of our lives in India acted itself in dâk-bungalows . . . so many men have died mad in dâk-bungalows." One can imagine that in Champawat, with its own volatile history, such structures were especially tainted by their overtly colonial implications.

The Tahsildar, upon hearing that a British tiger hunter had arrived in Champawat and repaired to the old dak bungalow, quickly scrambled up the dirt road to greet him, and to suggest that he might have better luck finding the tiger at another bungalow outside of town, near a village called Phungar. The tiger had recently attacked a number of people near Phungar, the Tahsildar insisted, although that may not have been the only reason for encouraging a relocation. Regardless, Jim Corbett, who appreciated the concern of the Tahsildar, agreed to move camp early the next morning to the bungalow outside of Champawat proper, and the Tahsildar, who appreciated Corbett's cooperativeness, agreed to meet him there for breakfast and do his best to help. And even if the Tahsildar did have misgivings about this British hunter's arrival, he surely recognized the importance of his mission. As a leader in the community, it was the Tahsildar's duty—just as it was Corbett's—to rid the town of the tiger by any means necessary. Collaborating with the British would have seemed a small price to pay for restoring normality to a region that had been terrorized for four years. And it is possible that the Tahsildar also sensed that this British hunter was somehow different; that in Corbett's strange grasp of the regional language, his bizarre intimacy with the animal world, his peculiar habit of smoking alone and mumbling to himself in the moonlight, the Tahsildar finally saw someone who might actually have a chance.

Someone just as unusual, just as unpredictable, just as *extraordinary* as the tiger itself.

CHAPTER 9

||||||||||||||||||||

AN AMBUSH IN THE MAKING

But what about the tiger? That elusive three-hundred-plus-pound female specimen of *Panthera tigris tigris* that could be heard roaring in the night, that pulled women out of trees and farmers from their fields? Where was it while Corbett scrambled to make sense of its lethal rounds and its deadly routine?

The fact that it killed humans may have been exceedingly unusual, but the manner in which the man-eater hunted and fed was definitely not. As the naturalist George Schaller notes in his seminal work, *The Deer and the Tiger*, "After the prey has been killed, the tiger usually drags or carries it to a secluded place, preferably into a thicket near water . . . The cat usually grasps the prey by the neck and drags the carcass between its forelegs or along the side of its body." According to Schaller, tigers are so powerful that animals as large as "a 400–500 pound buffalo, which three men find difficult to move, are readily pulled for several hundred feet by the cat." This explains the horrifying ease with which the Champawat Tiger carried off its victims into the forest; virtually all would have been under two hundred pounds, many considerably less than that. Schaller even writes of one tiger that was observed to have "jumped 15 feet up the bank of a stream while carrying a 150 pound carcass."

This is precisely the sort of behavior the Champawat exhibited upon yanking its Pali victim from the oak tree and scrambling up the steep side of the ravine.

A clear image of a tiger's eating habits can be formed from Schaller's observations involving its far more usual prey. After witnessing tigers dispatch numerous chital and gaur, he took note of how "the tiger begins to feed, regardless of the time of day, as soon as it has moved the carcass to a suitable locality." The tiger's carnassial teeth make excellent cutting tools, and "with a combination of cutting, pulling, and tearing, the cat rapidly bolts down the meat, skin, and viscera." The tiger generally begins with the rump and hindquarters, where the most flesh is to be found, and then slowly works its way up, feeding voraciously for up to an hour before stopping to rest. After a period of sleeping, grooming, and drinking— during which it sometimes hides or covers its prey—the tiger will return to continue its feeding, engaging in this cycle of resting and eating until the edible portions are entirely gone. The process can take several days, depending on the hunger of the tiger and the size of the prey. One tiger was observed to have consumed an entire 400-pound cow in just 4 days; a 250-pound barasingha took a different tiger 3 days to eat, while some tigers have been known to devour small-sized pigs or chital deer in just one sitting. With this in mind, a single tiger could readily consume an adult human in two to three days, which is precisely what the Champawat appears to have done in Pali.

It is likely that by the time Corbett found the tiger's feeding site in Pali almost a week after the initial attack, it had probably been abandoned for several days. And given the general lack of fastidiousness on the part of a famished tiger, it also becomes clear why so little remained to be taken back to the village for cremation. According to Schaller, "The tiger usually eats up its prey so completely that almost nothing remains for the scavengers." Skin, internal organs, and bones can all be ingested; some cats even eat hooves. With rough, papillae-covered tongues designed for shredding, and a row of sharp incisors made for snipping, fur, flesh, feathers—and in the case of humans, even clothing—can all be neatly stripped away and eaten or discarded as the cat sees fit. Tigers are nothing if not efficient, the product of millions of years of refinement. Efficient in how they hunt, efficient in how they kill, and efficient in how they feed. There is no malice or cruelty in what they do, any more than there is malice and cruelty in how a cow eats grass.

And it is in the interest of survival that tigers seldom stay long once the feeding is done. "When the last edible scraps of a kill have been devoured," writes Schaller, "the tiger usually leaves the site and either begins to hunt again or rests in another locality." As for the Champawat, it seems to have had little interest in the latter. It must have rested enough while Corbett was searching the pine groves around the village of Pali, desperately trying to find its tracks. A week had passed—roughly the time it takes for a well-fed tiger's hunger to return.

What's most interesting, however, is that the residents of Pali seemed to have had some notion of the tiger's eventual return to Champawat. That was, after all, where they advised Corbett to go. How did they know? Again, the answer may also lie in observed tiger behavior. James Inglis, a nineteenth-century chronicler of life on the Nepalese frontier, remarked how "like policemen, tigers stick to certain beats." In effect, many wild tigers have established routes they follow through their territory, doing their rounds while searching for food, with a centralized "home base" where they inevitably return and spend much of their time. The French tiger observer William Bazé supports this notion, claiming that the tiger "is very strongly attached to his permanent quarters and uses them as his base for his foraging expeditions." This behavior is also backed up by Schaller, who writes, "A tiger appears to have a center of activity within its range where it spends much of its time." And this "center of activity" can be long-standing: records exist of tigers staying in a particular locale for fourteen, fifteen, and even twenty years when food is plentiful. The phenomenon seems especially common in places where cattle-snatching is easiest—tigers with ready access to livestock are not especially inclined to move. But in the case of the Champawat, this tiger took up people-snatching instead.

Taking all of this behavior into consideration, a picture emerges of how the Champawat hunted. Our understanding is supported by the testimony of Nara Bahadur Bisht, the elderly friend of Peter Byrne, who remembered the tiger from his boyhood in Nepal. According to that account, the tiger was known to execute the majority of its kills around the town of Rupal—hence its Nepalese moniker, the Rupal Man-Eater. But as Bisht would recall, it also claimed victims in smaller villages scattered across the Nepalese district of Dadeldhura. When the tiger was finally driven out of Nepal and across the Sharda into India, it essentially adopted the same pattern of behavior, with Champawat as its new "home base." And when one looks at the two foci of its hunting, it makes perfect sense. Rupal and Champawat were both comparatively large, densely populated towns. Humans would have always been in relative abundance. Given that tigers are peripatetic within their own territory—a natural result of the instincts for both finding mates and fending off rivals—the Champawat did a regular "beat" of its own turf, traveling well-established routes across the middle hills. And when it came upon an easy target—a group of villagers collecting fodder or a young man tending to his fields—it took advantage of the opportunity to launch an attack. But as was the case in Pali, the village would immediately hunker down like a fort under siege, with the inhabitants refusing to leave the protection of their homes. Perhaps the tiger would linger for a few days, but without readily available food, it would inevitably return back to its base, the center of its territory where prey was abundant.

As Corbett began to realize, thanks to the villagers' accounts, although the tiger never did stop moving, its seemingly random array of attacks at far-flung locations were not quite as random as they had initially seemed. While the tiger's territory was large and its hunting grounds widely dispersed, it did appear to have a routine of sorts. Abandoning the small village of Pali, it circled back and returned to Champawat, possibly walking along dry streambeds during the day to avoid detection, and switching to established roads when darkness fell. A tiger actively searching for prey can cover a hefty amount of ground, traveling as much as twenty to thirty miles in a single night. At such a rate, it's quite possible that the tiger covered the distance between Pali and Champawat in one evening, although if it stopped to engage in the usual feline pursuits of marking its territory with urine and scent, drinking at streams, and resting during the sunnier hours of the afternoon, the journey may have taken several days. A *normal* tiger would have visited the places most likely to yield wild prey—watering holes where deer gathered, game trails where bushpigs foraged. This tiger, though, while similarly motivated, would have known through experience where its two-legged prey was certain to be found: on the outskirts of villages, on well-traveled roads, on the banks of clotted earth at the edge of tilled fields. Although a tiger's sense of smell is keen, particularly when it comes to detecting other tigers, it seldom uses it for hunting. Instead, tigers rely heavily on sound and sight, and after years of hunting people, the Champawat would have known what to keep its eyes and ears alert for: the babble of human voices, the clang of tools and pots, our gangly, upright gaits. And while tigers generally search out and stalk their victims, they also have been observed to hunt by hiding and waiting at places they know that their prey will soon be: an ambush, you might say. So when Corbett arrived in Champawat on May 9, 1907, it's possible—perhaps even likely—that the tiger was already there.

Hiding. And waiting.

CHAPTER 10

A LITERAL VALLEY OF DEATH

Jim Corbett's hunch seemed to have been correct. At breakfast with the Tahsildar early the next morning, having moved camp to the new bungalow outside of town, he watched as two men came scrambling frantically up the hill. Breathless, they announced that the tiger had just killed a cow in a village ten miles away.

Having already seen how efficiently the tiger had consumed its victim back in Pali, Corbett hurriedly gathered up his rifle, stuffing just three cartridges in his pocket. This was an old habit from his youth, when powder had been expensive and every bullet precious, but it was also a matter of practicality. Familiar with the speed and stealth of a tiger, Corbett knew that in the event of an encounter, a single shot was likely the best he could hope for. Either the tiger would flee or it would attack, with both scenarios offering little time for reloading. If death came for either of them, it was likely to be over in seconds. Corbett could only hope it would end in his favor. The Tahsildar wished him luck and promised that he would return in the evening to spend the night at the bungalow—assuming, of course, the hunter made it back alive. Corbett thanked him and set off with his guides, the three of them taking the packed-earth trail to the village at a blistering pace.* If it was indeed his tiger that had killed the cow, Corbett knew there was a fair chance that it was still close by, hovering near its prey.

If he was fast enough, he might still catch it.

The path was uneven and rutted but mostly downhill. The scattered stone abodes of the village materialized through the pines, as did the frantic farmer whose livestock had been killed. Corbett greeted him in Kumaoni and, rifle at the ready, asked to be taken to the carcass immediately. The farmer obliged, leading Corbett to a nearby cowshed, where the hunter would have easily detected, mingled with the usual scent of manure and hay, the unmistakable iron tang of freshly spilled blood.

Twisted and mangled in the corner was the body of a calf, already half eaten. And it had indeed been killed by a predatory cat. But no, as Corbett could tell right away, this was not the handiwork of a tiger. Even a cursory examination of the bite marks and tracks made it abundantly clear that this young cow had been killed by a leopard—an animal Corbett was on intimate terms with as well. He had killed his first leopard with a borrowed .450 Martini-Henry rifle while just a ten-year-old boy in Kaladhungi; the cat sprang at him in the forest while he was on a hunting expedition, and he acted instinctively, hitting its spotted hide in midair

---

*. One obvious question that does emerge from Corbett's account concerns his companions from Nainital, specifically: Why didn't they accompany him into the forest when seeking the tiger? The most obvious answer would seem to be that for the most part, thanks to the aforementioned gun laws, they were not allowed rifle permits, and would have had little training in the use of firearms. One exception may be Bahadur Khan, who is believed to have been among the six, and who would later go on to hunt with Jim Corbett, both as a gun-bearer and fellow shikari. At the time, however, Corbett seemed to have preferred taking only those with him who knew the lay of the land and could serve as guides; he considered it unwise, possibly even dangerous, to take along inexperienced hunters—at least until the tiger was located and beaters were needed.

and showering himself in blood. The young Jim Corbett was still so small at the time, he actually had to get his older sister Maggie to help him retrieve the dead leopard and carry it back to Arundel, the little stone cottage where his family made their home.

Leopards were relatively common in both the lowland *terai* and the middle hills of Kumaon, and they could be just as dangerous as tigers when they took to man-eating. Although considerably smaller in size—even the largest males were seldom over 170 pounds—they were more than capable of dispatching an adult human. One of the most infamous of such cats, known as the Leopard of Rudraprayag, would go on to kill as many as 125 people in the 1920s, most of them pilgrims traveling between the Hindu shrines of Kedarnath and Badrinath. In fact, man-eating leopards—though uncommon— were rumored to be even more fearless than man-eating tigers, and were known to break into homes at night and tear down walls to get their victims while they were sleeping.

This leopard, however, was evidently not a man-eater. Killing a cow, although financially onerous for a poor farmer, was not beyond the realm of normal leopard behavior. The year before, leopards had been responsible for the death of 2,744 head of cattle in Kumaon alone—almost twice the toll of 1,370 head attributed to tigers. Not surprisingly, leopards were generally classed as vermin by the colonial government. To Corbett, however, this particular leopard was a false lead and nothing more. He thanked the two men who had brought him there and gave them a few rupees for their trouble, but quickly set off back toward the bungalow.

Corbett arrived at the hut just before nightfall, disheartened to discover that the Tahsildar had not yet returned. The last of the sunlight was fading beyond the hills, and the surrounding valleys were beginning to pool with shadow; only a few minutes stood between him and the darkness. And once that darkness came, it would no longer be safe to venture outside beyond the bungalow's door. With nothing to do, Corbett felt anxious and uneasy—he didn't want to waste the last precious minutes of daylight. The *chowkidar,* or caretaker of the bungalow, must have noticed Corbett's frustration, and mentioned a nearby watering hole where he believed he had seen a tiger drinking.

Reshouldering his rifle once again, his expectations raised a second time, Corbett let the man guide him to the spring. But while there were indeed a few scattered animal tracks, he found no trace of the tiger he sought. He had studied the pugmarks of the Champawat closely at the kill site back in Pali, with a highly trained shikari eye. No, the man-eater hadn't been there—not recently.

The final disappointment came soon after the arrival of the Tahsildar, just as the night was beginning to fall. He listened with rapt attention to the stories of the day, only to inform Corbett that he could not stay at the bungalow with him as he had said he would. Perhaps the Tahsildar actually did have urgent business to attend to back in Champawat, or possibly he was concerned about appearing *too* friendly with the British—after all, finding the tiger depended on the cooperation of the local townspeople and villagers, and managing his relationship with the colonial government was a delicate affair. Regardless, after chatting affably for a few minutes in the gathering dusk, he apologized and said that he had no choice but to return to town—a change of plans that dismayed Corbett, as he was beginning to realize just how instrumental the Tahsildar might actually be. Corbett's growing admiration for the man is captured in his hunting memoir, *Man-Eaters of Kumaon*:

On returning to the bungalow I found the Tahsildar was back, and as we sat on the verandah I told him of my day's experience. Expressing regret at my having had to go so far on a wild-goose chase, he rose, saying that as he had a long way to go he must start at once. This announcement caused me no little surprise, for twice that day he had said he would stay the night with me. It was not the question of his staying the night that concerned me, but the risk he was taking; however, he was deaf to all my arguments and, as he stepped off the verandah into the dark night, with only one man following him carrying a smoky lantern which gave a mere glimmer of light, to do a walk of four miles in a locality in which men only moved in large parties in daylight, I took off my hat to a very brave man.

It appears to have affected Corbett greatly, that image: the Tahsildar, path lit only faintly by a single swinging lantern, gathering his robes and walking into the Kumaoni night, his pale form entering the darkness unarmed and unafraid. It was the sort of courage Corbett himself no doubt wished he could summon. He knew he was going to need it soon.

With the Tahsildar gone and his men already turned in, Corbett likely spent the evening on his own, eating a simple meal prepared by the *chowkidar,* smoking a procession of nervous cigarettes, listening in the dark to the collected whispers of the forest. Wondering where the tiger was at that moment. He understood the stakes. And if he succeeded in his mission, he knew what such a victory would entail. There would of course be the ancillary benefits: the plaudits from Charles Henry Berthoud back in Nainital, perhaps even recognition from the lieutenant governor himself—all of which could help pull a domiciled Irish lad out of a dead-end railway job at a backwater on the Ganges. But more important—much more, knowing where Corbett's allegiances lay— he would be saving scores of Kumaoni lives down the line. Corbett had learned from its tracks that the Champawat was an older tiger, though still in decent form—a fact that the Indian journal *The Pioneer* would confirm in an article published on June 7, 1907, stating that the cat "was not young . . . [but] in good condition." Which meant that even though the Champawat was past its prime, it still had years of killing before it. More gore-spattered topis, more claw-shredded saris, punctuating blood trails across stony, cold ra- vines. Corbett had seen it with his own eyes; he had wrapped the bone shards in funerary cotton with his own trembling hands. And he knew that although the final hunt would no doubt require assistance, he alone was in a position to stop the tiger.

If he failed, however—if the tiger came at him and his bullet did not hit its mark—there would also be consequences far more immediate, far more *personal,* as well. First, there would be the sheer force of the impact: a collision unimaginable, one that de- socketed his spine and split open his ribs. Then, the claws—ten of them, long as butcher blades, that stripped the flesh from his back and punctured his lungs. Followed at last by the teeth, a crushing quartet in the nape of his neck. And if his consciousness persisted beyond that, nothing remained but to be carried in the tiger's jaws as limp and helpless as a child, the sounds of civilization fading just as the wild chorus of the forest began, the hot rankness of its breath just a hint of the true horror to come . . .

It would have been enough to give anyone nightmares, and it seems even Jim Corbett was not immune to bad dreams. In his ac- count of the ordeal, he writes only of what transpired that night in the bungalow as something best left unwritten, a tale "beyond the laws of nature." And while the meaning of that phrase is not immediately clear, it seems reasonable to presume that he suffered some sort of panic attack or night terror, alone in the darkness, that left him sleepless and profoundly shaken. And who can blame him? In the inscrutable night, deep in hostile country, with a creature on its way that was believed to have killed nearly half a *thousand* people, terror was a sane reaction. With the bands of shadow turning to stripes in the cold glare of moonlight, and with dark shapes shuffling through the dust of the courtyard, it must have been difficult *not* to imagine, while gasping for breath and unable to move, that the Champawat was just outside, golden eyes watching, a Grendel come to collect its due. Years later, Corbett described such visions:

> Few of us, I imagine, have escaped that worst of all nightmares in which, while our limbs and vocal cords are paralysed with fear, some terrible beast in a monstrous form approaches to destroy us; the nightmare from which, sweating fear in every pore, we waken with a cry of thankfulness to Heaven that it was only a dream.

Only, for the people of Champawat, among whom Corbett now found himself, the monster that stalked them was no dream. Affirmation of that fact came the next morning after the arrival of the Tahsildar, who had kept his promise and returned to the bungalow as soon as he was able. Corbett, no doubt groggy after his harrowing night, was much relieved to see his friend arrive safely. The two men were talking, discussing the tiger's habits and attempting to presage its next move, when a runner from a nearby village suddenly appeared. The messenger came screaming up the side of the mountain, and in doing so, rendered their divinations all at once superfluous.

Come quickly, sahib, the man begged of Corbett, using the deferential form of address generally reserved for the British, *the man-eater has just killed a girl!**

It was a bittersweet reckoning. Another life had been stolen by the tiger's claws; but it meant there was a chance, if they were quick enough, of stopping the cat. The advice of the Pali villagers and the instincts of the hunter had both proven correct—the tiger had indeed returned to Champawat. It had killed again.

Corbett didn't waste a minute. He was already dressed in short pants and rubber-soled shoes—his outfit of choice when it came to pursuing large game. All that remained was a rifle. Informing the Tahsildar of his intentions as he raced past him, he ducked back into the bungalow to fetch a gun. Rather than selecting his old friend the .450 black-powder Martini-Henry rifle, or a lighter, more manageable .275, which he would come to rely on later in his life, Corbett chose from his private arsenal a double-barrel .500 modified cordite rifle, swapping out conventional black-powder cartridges for their more powerful nitro counterparts. Effectively, this hybrid weapon was a converted elephant gun, a new firearm-and-ammo combo capable of taking down everything from rhinoceros to Cape buffalo to the biggest of tuskers. High-powered double-

---

* The Champawat Tiger's 436th alleged victim also stands out as one of the only victims that may have been positively identified by historians. In 2014, Kamal Bisht, my guide in Kumaon, along with another group of Jim Corbett enthusiasts, located an elderly man in the village of Phungar, just outside of Champawat proper, who claimed that the girl had been the sister of his deceased father, and that she went by the name Premka Devi. He also believed that she was fourteen years of age when the attack occurred—which seems to match with Corbett's observation that she was a teenager, although he estimated her age to be sixteen or seventeen. Other details of the elder gentleman's recollections match Corbett's account as well, and although difficult to prove, the identification certainly does seem credible.

barrel rifles of this kind had been first developed for hunting in Africa and Asia in the mid-nineteenth century, when it became clear that neither muzzle-loaders nor single-barrel rifles were up to the challenge of stopping charging large game in its tracks. With the arrival of cordite propellant in the late nineteenth century, however, bullet velocities became attainable that had never been possible with the older, black-powder cartridges. In 1907, Corbett was straddling two eras—that of black powder and that of cordite—and by deciding to use the upgraded cordite cartridges in an oversized black-powder weapon to take on the Champawat, he made a strategic choice. As to why, exactly, he would choose this unwieldly twin-barrel shoulder cannon over something less awkward, the answer is simple: just like elephant and rhinoceros hunters in Africa, he had to stop the Champawat from charging. A smaller-caliber or lower-velocity weapon might have been easier to move through the jungle with, but it would have been far more likely to leave a wounded, enraged animal. In the case of a normal tiger, hunted from atop an elephant or raised machan, this would not have been so problematic—upon fleeing, a wounded tiger could be easily tracked and finished off later. But with a creature that had already attacked and devoured close to five hundred people, there was no guarantee that a hunter on foot would cause it to flee. An aggressive attack in defense of its kill seemed more likely, in which case blowing a pair of fist-sized holes through its striped hide was not an issue. After all, Corbett wasn't interested in acquiring a pristine tiger skin—he was trying to stop a serial killer.

With his rifle and customary three cartridges in hand—two for the twin barrels and one for an emergency—Corbett joined the Tahsildar and the messenger, and the three of them sprinted in silence down the hill. For a few minutes there would have been no noise but the sound of footfalls on packed earth and their own harried breathing.

When they arrived at the village, just a few miles outside of Champawat proper, a riot of yelling and pleading broke loose, as the frantic residents attempted to explain all at once what had occurred. One man among them was able to help calm the hysteria, and it was he who Corbett asked for a full report. Taking Corbett aside, he pointed to a scattering of oak trees just two hundred yards from the village. The story that followed was achingly familiar. A small group had been collecting firewood beneath the trees for their midday meal, when the tiger had materialized and come tearing into the group like a covey of quail. The others screamed and ran for their lives as a lone girl was engulfed by its stripes and claws—the tiger latched onto her neck before spiriting her away into the depths of the forest. The wife of Corbett's new confidant, who had been among the foraging party, pointed out the specific tree where the attack had occurred—she explained how they had not even seen or heard the tiger until the creature already had the poor girl in its jaws. The silence and suddenness of the attack are corroborated by the aforementioned June 7 article published in the *The Pioneer,* which describes how "about mid-day of the 11th some 25 women and girls were gathering leaves together when a tigress appeared, and, seizing a young girl, carried her off with hardly a sound." It was a tragedy that no one saw—or heard—coming.

Telling all gathered to stay where they were, Corbett checked on his rifle and started across the exposed fields to the kill site.

Arriving at the tree, Corbett was struck by the lay of the land. It was, in his own words, "quite open," and he found it "difficult to conceive how an animal the size of a tiger could have approached twelve people unseen, its presence not detected, until attention had been attracted by the choking sound made by the girl." Later in his career, Corbett would become more intimate with the incredible ways in which man-eaters downed their human prey, but the Champawat was the first of its kind he had encountered. Unlike wolves, for example, tigers virtually never hunt in packs; they are not especially long-winded and seldom chase prey over long distances. Their chief weapon, rather, is stealth—coupled with an astounding burst of speed. Belly flattened to the ground, creeping up in small increments on padded feet, they are capable of making themselves hidden in grass that's no more than knee-high.

Aiding them in their clandestine endeavors is one of the most effective sets of camouflage in the animal kingdom. The stripes of the tiger naturally break up its outline, blending seamlessly with the shadows of tall grasses and jungle leaves. Since most of its prey is color-blind, the tiger's orange fur is not noticed, and for those animals that can see color—including humans—its tawny hues meld extremely well with crepuscular light. This inherent stealth capability, when partnered with a leap radius that can reach thirty feet, and a top speed close to forty miles per hour, means that the eventual strike—when it finally comes, for tigers are also nothing if not patient—is almost always a surprise, and blindingly fast. And with their propensity for attacking from a rear angle, it's also very likely that in this case, the Champawat's human victim never even knew what hit her. Perhaps the faintest of rustling when it launched from the ground, a soft puff of displaced air from its careening body, all registered in the back of the mind at nearly the same moment nature's nearest equivalent to a short-range missile exploded upon its target.

And the evidence of that collision between predator and prey was right there beneath the oaks for Corbett to see. The exact spot where the girl had been killed was marked by a fresh pool of blood and a broken necklace of bright azure beads; the moment of her death, at least in Corbett's mind, was rendered starkly by that sickening contrast of crimson upon blue. No doubt swallowing down a lump of fear, and perhaps some inklings of nausea too, Corbett raised his rifle and followed the tiger's tracks, which were interrupted at steady intervals by splashes of blood where the girl's head had hung down from the tiger's mouth.

After half a mile of steady tracking, he came across the girl's sari, and at the top of the following hill, her skirt, both evidently ripped off by the tiger as it prepared to feed. From there, the drag marks took him to a thicket of blackthorn, from which strands of something long and dark were dangling and billowing from the stickers. Corbett stopped to examine one, puzzling over the peculiar moss, until he realized it was not moss at all. It was the girl's hair, which had snagged on the branches as the tiger passed.

Sickened and saddened by the sight, irrefutable evidence of the horror that had passed through the brambles only minutes before, Corbett was steeling himself for a prolonged pursuit into the brush when he heard it—the sound of feet, approaching fast from the rear.

There was no time to think, no time to plan. Time only for instincts, honed from all those years spent hunting in the jungles of Kaladhungi and the forests near Nainital. Corbett turned on a pivot, rifle at the ready, knowing one shot was all he would likely get off, and that he'd be lucky to get even that.

Alas, there was no tiger, and his fingers must have quivered for an instant before releasing whatever feeble amount of slack the twin triggers gave. Running up behind him was a man from the village, oblivious to just how close he had come to having two barrels unloaded upon him. He had been following right on the hunter's trail, his own beaten-up rifle swinging clumsily in his hands. Corbett's initial reaction was one of anger, as he had—in addition to almost blowing a local resident in half—explicitly asked that everyone stay put in the village until he had located the tiger. But his new companion, a patwari, or low-level village official, named Jaman Singh, informed him that the Tahsildar had sent him, as one of the few men in legal possession of a rifle, to go help the *sahib*—a gesture that Corbett apparently appreciated in spirit, no matter how misguided it may have seemed. Giving in, Corbett asked the village patwari to at least remove his heavy boots, which he feared would make too much noise in the forest, and that he keep his eyes behind them in case the Champawat circled back and attacked them from the rear. After all, two sets of eyes were better than one—particularly when a potential tiger ambush was involved.

Pushing on through the stinging nettles and scratching burrs, Corbett and his new companion followed the blood trail as it turned sharply to the left before plunging into a ravine choked with bracken and wild ringal bamboo. Then into a steep watercourse, scattered with loose stones and earth that the tiger had upturned on its way down—they were only seconds behind it now and they knew it. The watercourse became even steeper yet, cascading ever-downward and filling the ravine with a steady rush of falling water, amid which any number of muffled forest sounds could have been a lurking tiger. With the sheer rock walls rising on either side, they were easy targets should the tiger turn around—this they knew as well—and as the walls closed in the deeper they ventured into the gorge, the more perilous, perhaps even suicidal, their pursuit became.

The village patwari tugged on Corbett's sleeve on multiple occasions, his whispers wracked by sharp tremolos of fear, informing Corbett that he could hear the tiger, behind them, all around
them. It was becoming increasingly clear that the patwari, despite his possession of a rifle and admirable determination, was a
town dweller with very limited experience hunting, and that he
was proving to be more of a liability than an asset. Corbett had a
deep mistrust of guns in the hands of others, and an even firmer
unwillingness to put others in harm's way. Arriving at last at this
inevitable conclusion, he stopped at a steep stone spire, some thirty
feet high, and told his companion to climb the pinnacle of rock
and wait for him there. The patwari did so, and upon giving the
signal that he had arrived at the top, Corbett went on alone, his
thin, rubber-soled shoes angling and slipping over the wet rock as
he sidestepped and shuffled his way down the vertiginous ravine—
which, after a straight, steep shot of a hundred yards, ended at a
dark stone hollow with a pool of still water at its middle.

The tiger was gone, but traces of its feeding were not. Rifle
poised, ears pricked, Corbett inspected the site, discovering as
he did so a viscerally disturbing scene. Tigers usually feed near
water—that was not unusual—but the sight that met his eyes was
one that would be imprinted onto his memory:

The tigress had carried the girl straight down on this spot,
and my approach had disturbed her at her meal. Splinters of
bone were scattered round the deep pugmarks into which dis-
coloured water was slowly seeping, and at the edge of the pool
was an object which had puzzled me as I came down the watercourse, and which I now found was part of a human leg. In all
the subsequent years I have hunted man-eaters, I have not seen

anything as pitiful as that young comely leg—bitten off a little below the knee as clean as though severed by the stroke of an axe—out of which the warm blood was trickling.

Shaken, Corbett momentarily forgot the actual danger he was in. Disturbing the kill of a tiger is a perilous proposition, for if they have recently been feeding, they are seldom far away. But Corbett was, in his own words, "new to this game of man-eater hunting," and as of yet unprepared for all the hazards it entailed.

As he lowered his rifle and knelt to inspect the severed leg, a sudden sensation of being in extreme danger consumed him; a millisecond registration that something was out of place. Perhaps the faintest of rustling, or a soft puff of displaced air. But that was enough. Once again, Corbett acted on pure instinct. Never leaving his crouch, Corbett spun on his heels, ground the butt of his rifle against the earth, and put two fingers on the triggers.

Ears ringing from the blast of the double barrels, nostrils stinging with the acrid smell of cordite, Corbett blinked through the haze only to see that, rather than an enraged tiger, just a few clods of earth and loose sand had come tumbling down from the edge of a fifteen-foot bank directly above him. There was a soft stirring of ringal, the hollow bamboo stalks chiming ever so faintly, and the tiger was gone. Having not had time to properly aim, Corbett was confident he had not hit it. But that sudden and massive burst of gunfire at close range had been enough to discourage the ambush. The tiger had been on the cusp of a strike when his barrels roared.

Now it was the Champawat's turn to roar. Setting down what remained of the girl's body, the tiger let loose a thunderous cry. Corbett had just one bullet remaining, but that didn't deter him. Upon scrambling up the bank, he saw from a patch of bent Strobilanthes stalks where the Champawat had passed with its kill just seconds before. The terrain became increasingly difficult to cross, a welter of jagged rocks and deep crevasses. Despite the obstacles, however, Corbett was not far behind, following a trail marked clearly in blood. Leapfrogging over boulders, hopscotching across river stones as the tiger's growls echoed through the rocky abyss, Edward James Corbett was in a situation that must have felt strangely surreal—like the very worst of dreams. A slow-motion chase through a literal valley of death, after a striped creature that had in its maw its 436th human kill. It took many years for Corbett to fully collect his thoughts:

> I cannot expect you who read this at your fireside to appreciate my feelings at the time. The sound of the growling and the expectation of an attack terrified me at the same time as it gave me hope. If the tigress lost her temper sufficiently to launch an attack, it would not only give me an opportunity of accomplishing the object for which I had come, but it would enable me to get even with her for all the pain and suffering she had caused.

Bouncing from rock to rock along the bottom of the ravine, an awkward primate pursuing the single most lethal apex predator in the world, watching as, according to the article in *The Pioneer,* "portions of the body and its clothing were left behind," Corbett would have been unable to ignore the unpleasant truth that if the tiger were to decide to abandon its kill and turn around—and if his aim were to prove less than perfect—it would be his dismembered limbs scattered about the canyon floor. But after four grueling hours of pursuit, of seeing the thick clumps of rhododendron leaves stir just ahead, of hearing periodic growls rumble through the rocky passages, Corbett gave up. Night was closing in, and the bottom of a black ravine with a gleaming-eyed man-eater was the last place he wanted to be. As a tide of shadow began to seep into the valley, Corbett turned back and crawled his way out, pausing only to bury the severed leg of the poor young girl so her family could later retrieve it for cremation. Upon escaping the ravine, he found the patwari still there waiting for him at the top of the stone spire, much to the man's expressed relief—with the sound of the growls ringing through the valley, he had been certain Corbett had succumbed to the tiger. With their grim work done for the day, the two of them began the hike back to the village.

The patwari stopped at the place where he had hidden his boots, and while he struggled with the straps to put them back on, Corbett sat and had a smoke. With the distant peaks of the Himalayas catching the last of the day's light, reflecting it back across the foothills in a lambent gold, he studied the lay of the land and considered his next move. He knew going back down into the ravine alone was hopeless—the tiger clearly had the advantage in that rough and densely wooded terrain. One near-death encounter and four hours of hopeless chase had proved that. However, in the rippling landscape that spread out before him, Corbett saw an opportunity; a "great amphitheatre of hills," as he would call it, with a stream forming a narrow gorge that cut west to east, and with one especially precipitous hill directly opposite. The tiger would almost certainly stay with its kill to continue feeding—which meant that it wouldn't stray far for another day or two, at least. This tiger obviously preferred the low ground, retreating to the bottom of steep ravines that men could not easily reach, which was in part how it had evaded hunters for so long. It occurred to Corbett that if he could get enough helpers to man the length of the ridge from the stream to the hill, and then somehow manage to drive the tiger out from its quarters below, its natural line of retreat would send it out of its ravine and right into the narrow gorge that bisected the amphitheater. Where, if Jim Corbett's incipient plan worked, he would be waiting with both rifle barrels cocked. Unlike the tiger's present hideout deep in the ravine, this long, winding second gorge was relatively clear, free of foliage and rocks, and seemed to offer the only possibility of an unobstructed shot.

So that was it. There would have to be a beat. A line of men almost a mile long, all working together, marching into the brush to flush the tiger out into the open, while the hunter waited to spring the trap. It was roughly the same technique that had been tried in Nepal some four years before, and while it had succeeded in driving the tiger out of the country, it had also failed to capture or kill it. How much of this Corbett was aware of is difficult to say, although he did know such a beat would not be easy to replicate, particularly without trained elephants or experienced shikaris. It was to be, by his own admission, "a very difficult beat, for the steep hillside facing north, on which I had left the tigress, was densely wooded and roughly three-quarters of a mile long and half-a-mile wide." Difficult, yes, but not impossible. He knew that if he could get everything organized correctly and have the beaters follow his directions, there was at least a "reasonable chance" of him getting a shot.

All he had to do then was convince several hundred men, none of whom had ever engaged in a large-scale tiger hunt before, and all of whom hailed from a region famous for its distrust of colonial authority, to put their faith in an outsider and walk unarmed and helpless into a monster's lair.

He was going to need help.

||||||||||||||||||||||||||||||||

Corbett's first stop before returning to his bungalow was the little cluster of slate-shingled farmhouses that formed the village outside Champawat where the latest victim had been killed—and where, as it just so happened, the Tahsildar was already waiting. With the peach glow of sunset fading to the plum tones of twilight, the sore and limping Corbett must have been at once both nervous and relieved to see the Tahsildar—the closest thing in Champawat he had to a friend—cloaked and turbaned in dusky silhouette. Upon greeting him, perhaps Corbett related the horrendous events of the day, or maybe the exhaustion in his frame told the story for him. Smoking quietly together in the last snatches of daylight, Corbett must have finally worked up the courage to tell the Tahsildar of his plan. One can imagine the hesitancy in his voice, the uncertainty, because for once a representative of the colonial government was not giving an order so much as begging a favor—and not in English, as was customary, but in the native Kumaoni. Between puffs, perhaps, or even following a long draw, with the glowing butt held waist-high between his fingers, Corbett would have revealed the trap he had in mind, as well as his desperate need to get the people of Champawat on his side—something he feared might be impossible.

And although Corbett may not have seen it in the last of the gloaming, the Tahsildar, a man who no doubt remembered the tales of the old days, before the rebellion, before "The Mutiny," as the English called it, surely smiled. A sly, knowing smile, one that lit up his eyes and creased his hill-born face. A flick of the butt, a cartwheeling ember, vanishing into the shadows of the fallow field beside them, and then the Tahsildar was gone, rustling away through the grass on his way back to Champawat, racing the darkness, one step ahead of the night.

||||||||||||||||||||

CHAPTER 11

CONFRONTING THE BEAST

Early the next morning, Corbett emerged from the bungalow into a lingering darkness. The deep indigo of twilight still clung to the hillsides, although the first peaks of the Himalayas were beginning to blush. He had slept well, without incident. He left the threshold feeling refreshed and anxious to start the day.

Aside from his six companions from Nainital, and perhaps a few of the twenty men he had met on the road to Champawat, Corbett couldn't count on much in terms of assistance. Taking part in a beat was a hazardous affair—volunteers were rare, as no one was safe in the event of a mishap. The hazards of the *bagh shikar* are well documented in the colonial accounts of the era, with cornered tigers frequently charging the beaters, in some cases even dragging hunters down off of their elephants. Not surprisingly, many of these accounts, which are almost unanimously British in provenance and rich in hubris, frequently portray the Indian beaters—who were generally unarmed and coerced, sometimes forcibly, into taking part—as cowardly and ineffectual. Upon finding a tiger, the beaters are usually depicted fleeing from the forest "like so many rabbits from a warren when the weasel or ferret has entered their burrow," as the nineteenth-century tiger hunter James Inglis would claim.

Some hunters, like Inglis, even found such incidents amusing, as demonstrated in this passage, tinged with sadism, from his 1892 memoir, *Tent Life in Tigerland*:

> The beaters came pouring out of the jungle by twos and threes, like the frightened inhabitants of some hive or ant-heap. Some in their hurry came tumbling out headlong, others with their faces turned backwards to see if anything was in pursuit of them, got entangled in the reeds, and fell prone on their hands and knees . . . I, who had witnessed the episode, could not help . . . a resounding peal of laughter.

Amusing, perhaps, to a well-armed, well-fed, well-heeled English sportsman watching at a distance from the relative safety of an elephant back. Needless to say, an impoverished farmer being forced to walk into a forbidden forest, totally unarmed, to confront an enraged tiger, probably would not have found much to laugh at. Quite the opposite. Unlike the Shah kings of Nepal, who had carefully cultivated their hunting partnership with the Tharu, rewarding loyal beaters and elephant handlers with monetary gifts and *lal mohar* land grants, the British in India—and to an extent the subsequent Rana dynasty in Nepal—relied heavily on "volunteer" beaters who were in actuality anything but. Participating in a beat was a civic duty, akin to being drafted for military service, and it was often greeted with similar levels of disdain and resentment. At best, it meant a long, hard day of hacking through the densest of jungles; at worst, being fatally mauled by a five-hundred-pound beast.

Jim Corbett surely knew this. The locals were not foolhardy or cavalier, and they certainly did not owe him anything. As Corbett paced and smoked at the base of the tree where they had agreed to meet come morning—the same tree beneath which he had found the girl's bloody necklace the day before—he must have known that his chances were slim of organizing an effective beat, with or without the Tahsildar's help. In Corbett's own words, "That he would have a hard time in collecting the men I had no doubt, for the fear of the man-eater had sunk deep into the countryside and more than a mild persuasion would be needed to make the men leave the shelter of their homes."

And it seemed Corbett was right. At ten o'clock, the Tahsildar, true to his word, did appear, but with only one man at his side. Corbett surely greeted him warmly, in appreciation of the brave gesture, and he would have done his best to hide his disappointment. Perhaps he accepted, with the dark humor of a soldier on a hopeless mission, the inevitability of their impending failure: a small pack of men against an animal that had already killed so many more.

But within minutes, another pair of men showed up. Then another after that. And then, even more men still, slowly trickling down from the hills in twos and threes. One can easily imagine Corbett's pleasant surprise as their numbers reached, by his own count, 299 strong by the middle of the day.

It soon became clear how the Tahsildar had accomplished the seemingly impossible. In his wisdom and understanding, he had promised the inhabitants of Champawat that in this one instance, all forms of weapons would be allowed—that the "powers that be" would conveniently look the other way. Essentially, he repealed, at least in one town, for one day, the Indian Arms Act. And for the first time since the Rebellion of 1857, the population was openly brandishing an arsenal of weapons that, according to Corbett, "would have stocked a museum." Granted, most of their weaponry was severely outdated, and in bad condition after spending decades buried or hidden away. It was intimidating nevertheless.

Corbett stared in awe at the previously inconceivable sight of Kumaoni men carrying "guns, axes, rusty swords, and spears"— weapons that hadn't tasted blood in half a century, and that under normal circumstances would have landed any of their owners in a colonial jail cell. Fifty years after the downfall of Kalu Mahara, the people of Champawat had once again risen up and formed an army, only this time, their enemy wasn't the occupying British government, but rather the beast its indiscretions had unleashed upon them. Many had lost loved ones to the creature. At least one, as Corbett was soon to learn, had lost both of his sons and his wife—essentially, his entire family, wiped out by a single man-eating tiger. This was to be the day in which scores were settled; the men were armed and hungry for blood. And while Corbett assessed the scene, the Tahsildar simply loaded his own double-barrel shotgun with contraband shells and distributed ammunition to all who needed it.

Jim Corbett gathered the band of three hundred men around him, and with the help of the Tahsildar, explained how the beat would work: they would form a line along the edge of the ravine, directly above where the tiger was feeding. While they spread themselves out across the ridge, he would take his post beneath a pine tree directly across from it, which was easily identifiable, thanks to a lightning strike that had stripped its bark. When the signal was given—a wave of a handkerchief on the part of Corbett—those among the posse who possessed firearms were to fire them into the air, while the others beat drums, rattled gongs, rolled rocks— essentially anything that could disturb the invisible tiger down below. Shortly after which, Corbett intended to take a hidden position at the mouth of the gorge, waiting for the man-eater to come exploding from its rocky maw, enraged, confused, and ready to defend itself against anything that stood in its way. If it all worked, that final showdown in the clearing was where it would end. Either a man or a tiger would live to walk away.

The assembled men approved of the plan and quickly set off for the ridge, spacing themselves evenly to cover its full length. Corbett turned to circle back and take his post beneath the blasted pine, but as he did so, the Tahsildar stopped him. *I should come with you*, he said firmly. He had a hunch that this British shikari, regardless of his skill or experience, might require assistance.

The two of them went racing across the upper end of the valley, first clambering up the ridge of the opposite slope, then sidestepping halfway down its steep face to the twisted form of the dead pine. Here, the Tahsildar called a momentary halt. Unlike Corbett, who was wearing comfortable rubber-soled shoes—the sneakers of his day—the Tahsildar had a thin pair of leather brogans better suited for the dirt roads in town than the rocky faces of Kumaoni hills. They stopped only for a moment, for him to adjust his footwear, but that short delay was enough to disquiet the restless beaters spread out along the ridge. Assuming Corbett had forgotten to give the signal, the anxious townsfolk decided to start the beat on their own. All at once, the din ensued, the sound of some three hundred men firing rifles, pounding drums, and screaming as loud as their lungs would allow. Boulders were sent crashing down into the hidden depths of the ravine, and spears hurled blindly into the darkness below.

There wasn't a second to spare. Corbett unslung his rifle and made a skittering descent to the clearing at the mouth of the gorge,

150 yards away. He had no time to find a perfect blind—he spotted a stand of tall grass with a clear view of the gorge's black mouth, and he knew it would have to do. If the tiger attacked, he would be left exposed and vulnerable. Another ravine opened up behind him and to the left—he assumed that once the tiger broke free, it would cross the brief patch of open ground as quickly as possible and make for its shelter.

He would have only a few seconds to take his shot, at a moving target traveling at the speed of a Thoroughbred.

Corbett knew what to expect—he had been envisioning it for hours—but no amount of anticipation could have prepared him for the tiger's actual arrival. And he knew tigers were fast—quick enough to catch chital stags, even swamp deer—but exactly how fast, he had not realized until it broke through the trees. Amid the cries of the beaters and the clatter of their guns, the creature finally appeared, exiting the gorge higher than he had anticipated, and flying in his direction at a downward angle. A striped apparition, too fleet to be real, erupting from the shadows and streaking across a clear slope some three hundred yards away.

For a moment, all else must have lost its significance; the roar of the beaters deadening to a muffled hum, the thunder of the rifles dampening to a pale crackle. And cutting through that still and perfect moment in a liquid throb of movement, the Champawat Man-Eater. The deadliest animal in all of India, probably the world, cutting a furious swathe through the tall grass toward him, ears flattened, teeth bared. The same animal he had heard spoken of in hushed tones in Nainital tea parlors, suddenly taking shape; the same creature that had kept him and his men huddled in terror around a campfire, suddenly given form. Perhaps, for a moment, his whole purpose was called into question. Perhaps, just like the unknown boy in Nepal who first took aim and fired from his rickety machan, the notion of shooting this thing before him felt absurd, impossibly bold and monumental in scope, as if he were not killing a mere animal, but assassinating a king or a deity.

Perhaps—but if Corbett hesitated for a moment, the Tahsildar did not. From his flank post beneath the pine tree farther up the slope, he let loose both barrels of his shotgun in the tiger's direction, sending a duet of slugs screaming toward its hide. The shots missed, which isn't surprising given the distance to the target and the poor condition of his shotgun. But the sound of the report and the explosion of earth at its feet were enough to bring the advancing tiger to a grass-shredding halt.

Corbett now had a chance at a shot, albeit a long one. As the tiger reared and turned tail, he was able to get off a single "despairing" shot from his rifle. But this bullet missed as well, sending up another geyser of ruptured sod, and Corbett could only watch as the tiger vanished back into the thick undergrowth of the ravine. And he must have both praised and cursed his luck at once—he was still alive, but then again, so was the Champawat. His guess was that now, it would attempt to flee the gorge in the *other* direction, eventually racing up the steep side and into direct confrontation with the line of beaters—and who knew what carnage that might mean—or, it would simply hole up at the bottom and wait until darkness, in which case they would likely lose it for good.

Luck, however, is a funny thing. Even that which appears to have soured can take a turn quickly for the better. While Corbett stared into the hopeless ravine, dejectedly changing his spent cartridge and anticipating the screams of wounded men, something unexpected occurred. The army of beaters, having heard the gunshots, assumed that the Champawat had at last been killed. A sweeping roar of approval moved across the lip of the chasm, and all of the remaining ammunition was fired into the air in celebration. The collective tumult of that misguided hurrah was enough to jar the tiger loose from the ravine a second time, back toward Corbett.

It reappeared, showing its stripes once again, conjured on command as if by a Tharu *gurau*. The tiger traveled directly across the open bottom of the valley, crossing the stream at its base in a single, jaw-dropping leap, then headed straight for the cover of the next wooded ravine.

This time, Corbett was ready.

‖‖‖‖‖‖‖‖‖‖‖‖‖‖‖‖‖‖‖‖‖‖‖‖‖‖‖

The .500 modified cordite rifle came up in a single fluid movement. Corbett was aware that the gun, sighted at sea level, tended to shoot high at altitude, but he adjusted accordingly, getting a steady bead on the tiger then squeezing the first trigger.

This time, the shot struck home. A little farther back than he would have liked, but the Champawat had been hit. Snarling, its head bowed in pain, the cat spun around in the direction of the report, although Corbett was still hidden from sight by the patch of tall grass. The tiger's reflexive contortion gave him the cleanest shot yet, from not more than thirty yards away—a distance an attacking tiger could cover in less than two seconds.

With a steady hand, he took the slack out of the second trigger, the blotted orange of the tiger hovering at the cusp of his sight.

A second blast resounded through the valley. The bullet tore through the Champawat's shoulder, into its chest. Fully enraged, it bared its teeth and lowered its ears, prepared for a charge, searching wildly for the source of its agony. With Corbett still out of sight, it attacked the closest object it could find—a bush, standing across the stream, fastened at the joint to a protruding shelf of rock. Still dazed by the unexpected wounds, it bounded back across the stream and mounted the boulder, where it commenced in its fury to tear the bush to shreds.

Jim Corbett realized he had a serious problem. With a sinking sensation, like the nadir of a nightmare, his rifle became useless in his hands. His old habit from boyhood had finally caught up with him; he had spent all three of his precious cartridges. The Champawat was only feet away, in an absolute frenzy, and he was helpless to do anything about it. It was distracted for the time being, but Corbett knew that a seriously wounded tiger was the most dangerous kind of all.

The Tahsildar . . . Corbett shouted in his direction, toward his post higher up beside the blasted pine. His words were lost, however, in the mayhem of growls and the distant yelling of the beaters. The Tahsildar shouted something back, but his words were a jumble as well. And then, Corbett knew his only chance was to run to the Tahsildar and get his shotgun. Which would mean breaking cover from his hiding place in the tall grass and making a dead sprint across more than a hundred yards of exposed valley— hopefully before the Champawat had time to take note.

When describing the scene almost forty years later in *Man-Eaters of Kumaon*, Jim Corbett recounts what transpired summarily and succinctly, in the calm and unhurried tones of a lifelong hunter. But it's hard *not* to imagine, however, the feeling of raw, helpless exposure he must have experienced when he dropped his empty rifle and began running unarmed toward the pine, uncertain if the tiger was behind him or not. Every thudding footfall bringing him one step closer to the gnarled tree, that fixed goal bobbing on the bank of the hill. The Tahsildar stands, at first confused, then understanding at last what this puzzling Britisher from Nainital intends to do. He nods, a gesture pregnant with trust and meaning, and he tosses the running Corbett his only weapon. The old shotgun hangs poised for an instant in the trembling blue air—and Corbett grabs it, never breaking stride, grinding his heels into the earth and turning back down toward the stream.

And now there is the tiger: slowly turning its head from the shredded bush, and seeing the true source of its pain, indeed, of its lifelong misery, coming full speed down the opposite slope toward it. For a moment their eyes meet, and there is an understanding. A negotiation between hunters as old as time.

Corbett takes the stream in a running leap. His lungs are spent, he's gagging on fear. As he approaches the rock ledge, the tiger comes out to meet him, covering the stone shelf in long, loping strides. It stops at the brink, and so does he.

Corbett is face-to-face with the Champawat, clutching a slap-dash shotgun not so different from his boyhood muzzle-loader held together with wire. For a split second, he must have been that young boy again, fatherless and lost and alone in the jungle, frozen before a pair of golden eyes staring out from a plum bush.

The Champawat turns its fury upon him. Though severely wounded, perhaps even mortally so, it has enough fight left in it for one more kill. Gushing blood, hurling roars, it comes back to the rock ledge before him. It is only twenty feet away now, a single leap from having its jaws on his throat. It sees Corbett, clearly at last, and opens its maw to voice its rage, exposing its wounded mandible, its shattered teeth. And all at once Jim Corbett understands what's been done to this poor creature, a story written in malice and pain. But the number 436 leaves no room for pity, and twenty feet affords him no chance at escape.

The old shotgun comes up, cocked and ready, loaded with the last of the Tahsildar's cylindrical-slug shells. This is how it has to be. Given the weapon's poor condition, Corbett isn't even certain it will fire. It will work or he will die—it is that simple. The tiger crouches and prepares for its attack.

Corbett takes in a breath and hooks his finger around the trigger. Perhaps he hesitates, for the thinnest sliver of an instant. Once the trigger is pulled, two paths await him: extinction or transformation. Perhaps he already knows the future, as inevitable as the monsoon rains, as certain as the snows that cap the Himalayas. He closes his eyes to the gathering horror before him and he prays . . .

There is a single clap of thunder.

A pair of golden eyes goes dim.

One life ends, and another begins.

||||||||||||||||||||

CHAPTER 12

A MOMENT OF SILENCE

Between the final blast of the shotgun and the arrival of the beaters, there was contemplative silence. A calm minute filled with both satisfaction and regret. For while Corbett was content to have accomplished that which was deemed all but impossible, he was also deeply shaken—as he would be for the rest of his life—by the act of killing a tiger. He would later describe the peculiar sensation that followed the destruction of a man-eater as "a breathless feeling—due possibly as much to fear as to excitement—and a desire for a little rest." The Tahsildar joined him, and the two men stood vigil together. Just above them, the Champawat's head hung limp over the edge of the rock, releasing a slow drip of blood that dimpled the dust at their feet.

Once Corbett's nerves settled, he mounted the steep bank of the stream and approached the ledge to inspect the dead tiger. Just as he reached its limp body, however, the first of the beaters burst through from the forest, brandishing their assortment of guns and spears, whipped into a frenzy by the sight of the striped form sprawled out across the boulder. Few among them had not lost a loved one to its claws. The man whose entire family had been consumed by the cat was especially intent on tearing it to pieces, shrieking at the top of his lungs, "This is the *shaitan* that killed my wife and my two sons!"

With the help of the Tahsildar and some desperate pleading, Corbett was at last able to subdue the crowd, and the men's rage slowly shifted to a morbid sort of curiosity. One by one, they climbed to the protruding slab of rock and gazed at the Champawat up close, suddenly so much smaller and less imposing than it had been when alive. Glazed eyes, blood-matted fur, lolling tongue—it was almost pitiable, as if the *shaitan* that possessed it finally had been exorcised, leaving behind only the limp remains of its discarded vehicle. The men lowered the tiger to the ground so Corbett could inspect it more closely, and confirm what he had suspected at first roar: both the upper and lower canine teeth on the right side of the tiger's mouth had been damaged long ago, cut cleanly by a misguided bullet. The upper tooth was shorn in half, and the lower tooth, broken off all the way down to the bone.

The skinning of the tiger was delayed, though—the men asked to wait until nightfall, so that they could carry it through the surrounding villages and prove to their families that the beast was actually dead. The population needed to know that they could tend their fields without fear and walk on the roads without dread; that the night was theirs again. Corbett watched as the men lashed the Champawat to a pair of stripped saplings with their unwound dhoti cloths and turbans, formed a human chain up the face of the mountain, and passed the burden of its body up along to its crest, singing an ancient hill song in unison as they did so. This procession had not been seen in half a century: Kumaoni men celebrating the defeat of an enemy that they themselves had banded together to vanquish. In the tiger's demise, something seemingly irretrievable had been redeemed; something once lost had again been found. And now the Tahsildar joined them, as the men carried him up the mountain atop their broad shoulders, a symbol of respect, a gesture of triumph.

Corbett climbed the slope by himself, watching with a pang in his heart a celebration he knew he could not participate in. Still, it was moving to behold, and at the top of the valley, the Tahsildar rejoined him, unwilling perhaps to leave him totally alone. As the tiger was marched eastward by the assorted villagers along the ridge of the amphitheater where the killing had taken place, the two men took the road back to Champawat together. As they marched, they noticed a single ribbon of white smoke unspooling its way toward the heavens from the valley below. It was a funeral pyre. The family of the tiger's last victim had at last been able to recover her remains from the ravine, and now, at last, they were sending her home.

|||||||||||||||||||||||||||||||

The festivities went long into the night. From a courtyard in Champawat, Corbett and the Tahsildar watched a procession of pine torches wind through the valley, as the jubilant songs of the people rang out through the still air. Gradually, the procession made its way down from the surrounding hills and arrived at the Tahsildar's door, where the body of the Champawat was laid at Corbett's feet to be skinned at last. It was customary for the shikari who had fired the lethal shot to keep the head and skin as a trophy, and Corbett obliged. He was happy to donate the body to the local villagers and townsfolk, who believed that lockets containing pieces of the tiger could serve as powerful talismans, and protect their children from future attacks. As Corbett crouched beside the tiger's body, running his knife along its hide, he noticed that the final shot, the coup de grâce he had delivered with the Tahsildar's battered shotgun, had actually struck the Champawat in its foot—not its roaring maw. But that had been enough to bring the creature down.

By the time Corbett had removed the tiger's skin, the town elders had planned the feast that was to be held the next day, to celebrate the end of their four-year ordeal. But it was a celebration Corbett either would not—or could not—attend. His railway job back on the Ganges was beckoning, Nainital was still seventy-five miles away, and perhaps he knew that while his presence at the celebration would certainly be tolerated, even welcomed, it was again a victory that was not his to celebrate.

As the last of the stragglers left the courtyard, elated at being able to travel the roads at night without fear once again, Corbett had one final smoke with the Tahsildar. Between thoughtful drags on their cigarettes, enjoyed beneath the Kumaoni stars, Corbett told his friend that he could not stay for the celebration—that the Tahsildar would have to take his place at the head of the table instead, and that he deserved it. The Tahsildar, who seems to have understood far more than Corbett lets on in his writing, may have smiled, and told him that he would be honored.

||||||||||||||||||||||||||

Jim Corbett left Champawat at sunrise the next day on a borrowed horse, leaving early, riding alone, the rolled skin of the Champawat strapped to the saddle. He told his men from Nainital that he would meet up with them again at Devidhura, where he intended to spend more time cleaning the hide. On the way, however, it occurred to him that there was someone in Pali—the village where all of this had begun—who might be interested in seeing the Champawat for herself. Corbett found the stone farmhouse beside the village, dismounted from his horse, and laid out the skin of the Champawat in front of the family of the girl that the tiger had killed the year before. The surviving sister of the victim, who had been too traumatized to even speak of the incident before that day, called excitedly for all the village to come and see what the *sahib* had brought. Corbett drank tea with the people of Pali and recounted the details of the hunt, telling them of the bravery of the men who had helped kill the tiger, not in English, but in Kumaoni.

And it was a sentiment that he seemed to have echoed in his official report to the colonial government, upon his return to British society. Several months after the Champawat was killed, Sir John Hewett, the lieutenant governor of the United Provinces, hosted a special durbar reception in Nainital to commemorate the event—a ceremony at which both the Tahsildar and the patwari were presented with an engraved rifle and knife, respectively, as tokens of the government's gratitude. Of course, Jim Corbett doesn't mention the following in his account—to do so would have struck him as boastful and crass—but the lieutenant governor made sure to present him with an engraved rifle as well, a sleek and modern bolt-action .275 Rigby. It was a vast improvement over his old double-barrel black-powder rifle, and a prescient gift. Because as the people of Kumaon were soon to discover, the death of the Champawat had not marked the end of anything when it came to man-eaters.

Far from it.

CHAPTER 13

AN UNLIKELY SAVIOR

This book began with the premise that the Champawat Tiger was not a freak of nature, but rather a man-made disaster. That its arrival at the very cusp of the twentieth century was not an isolated aberrance on the part of the animal kingdom, but the direct result of decades of environmental mismanagement on behalf of the governments of Nepal and India; a catastrophe at least half a century in the making, arguably considerably longer. And if this premise holds true—if the ecosystems of the region had degraded to the point where apex predators could no longer subsist on their natural diet in a wild habitat—it only stands to reason that the death of the Champawat should have marked the beginning of an era of unprecedented man-eaters, certainly not the end.

A harbinger, you might even say.

And this was precisely what occurred. Prior to the Champawat, as even Jim Corbett himself admitted, man-eaters in the divisions of Kumaon and Garhwal were all but unknown—tiger fatalities were perennially in the low single-digits, if that. The British authorities barely even bothered to keep statistics, and the few attacks that did warrant mention were usually the result of startled tigers acting defensively, protecting either their territory or their young.

In the Champawat's wake, however, a veritable cavalcade of serial man-eaters suddenly appeared in Kumaon, seemingly compelled by the same combination of diminished habitat, reduced ungulate populations, and human-inflicted injuries. And it was upon this plague of man-eating tigers and leopards that the legend of Jim Corbett was built.

In killing the Champawat, Corbett had made a name for himself with the colonial authorities. And as an ever-increasing number of aggressive cats emerged from their ruined forests to seek human prey on the edges of villages and towns, Corbett was inevitably summoned, like some mythological hero, to dispatch them one after the other. Among the most famous of the man-eaters he hunted, there was the Leopard of Panar, a cat that was reported to have claimed almost as many victims as the Champawat, and which he finally terminated in 1910. Later came the aforementioned Leopard of Rudraprayag, killer of 125 humans, which he stalked and killed in 1926. Then, the Chowgarh Tigers, a trio of cats that Corbett was called upon to hunt in 1929. Next, the Mohan Man-Eater, another tiger, which Corbett shot in 1931. There was the Kanda Man-Eater, killed in 1933; the Chuka Man-Eater of 1937; and finally, the Thak Man-Eater, which he finished off in 1938, at the age of sixty-three.

And these are just the most celebrated hunts. In total, Jim Corbett is credited with having ended the sprees of more than thirty problem tigers and leopards in the region, with a total estimated human victim count that may have exceeded one thousand. The reality of the hunts, however, despite the occasional moment of heart-pounding excitement, was in most cases far from romantic—Corbett was quite forthcoming about this when recounting the many cold, damp, and fruitless nights he spent shivering in treetop machans or trying not to fall asleep beside baited gin traps. But his persistence and his knowledge of the region's flora and fauna usually paid off, and he bagged his man-eater. It was thanks to his early successes—beginning, of course, with the Champawat Tiger—that Corbett became the British government's go-to man when it came to dealing with the man-eaters that began cropping up in the divisions of Garhwal and Kumaon. He would retain his post at the railway ferry terminal in Mokameh Ghat for several more years, but whenever the need arose, he would go on leave to hunt down the wanted predators. And as one might expect of a man who could stop ferocious man-eating beasts that no one else could, Jim Corbett gained a certain celebrity.

It was a notoriety that Jim Corbett, though self-effacing by nature, seemed to have accepted, if not actively cultivated. As his legend grew, the "Gentleman Hunter" from Nainital was increasingly invited to visit the rarefied air of social clubs and formal events, becoming a frequent (and no doubt entertaining) dinner guest of the colonial elites. His assistance was requested when the government organized elaborate tiger hunts for prominent officials and visiting celebrities—including the Viceroy of India himself, Lord Linlithgow, who befriended Corbett and became a patron of sorts. Even the Queen of England, Elizabeth II, would eventually cross paths with this domiciled son of an Irish postman. Corbett, along with some aristocratic friends, would host the soon-to-be monarch when she stayed at the Treetops Lodge during a visit to Kenya just prior to her coronation—an event that would become a memorable high point in the old hunter's career.

The greatest boost to Corbett's fame, however, came in the form of his hunting memoirs, the most famous of which was *Man-Eaters of Kumaon*, first published in 1944 by Oxford University Press.

The book became a major bestseller, praised for both its accuracy and its storytelling, and by 1946, it had already sold more than half a million copies. For the American release, a lavish reception was held in New York City at the Pierre hotel. Corbett was not able to make the book signing—during the war, he had taught jungle tactics to young recruits in India, and the malaria he had acquired there had not left him in the finest of fettle. In lieu of the author, however, his publisher flew in captive tiger cubs on a private plane to sign books in fluorescent ink with their paws. All of which the American literary press ate up with a spoon. Riding the tide of the book's popularity, other bestselling Corbett titles would follow, including *The Man-eating Leopard of Rudraprayag, My India,* and *Jungle Lore.* The success came as a pleasant surprise to Corbett, who had never anticipated a career as a writer. He did make sizable donations with the proceeds, but he was also able to live comfortably in his later years—and support his sister Maggie—thanks to the royalties and fame that the books provided. It was an existence far removed from his hardscrabble youth, when his mother had been forced to rent out half their home to pay off her dead husband's debts, and necessity had compelled him to scour the forests for bush meat.

Effectively, Jim Corbett had achieved the life of a true colonial *sahib,* complete with high-society friends, a new tenanted estate in Kaladhungi, and a long list of titles and accolades attached to his name. These included an honorary magistrateship, a Volunteer Officers' Decoration for his service in two world wars, the Kaisar-i-Hind gold medal, and both the OBE (Order of the British Empire) and the CIE (Companion of the Indian Empire), the latter of which was awarded to him at the Birthday Honours of

King George VI. Back in India, Corbett was given an exceedingly rare honor known as "The Freedom of the Forest." This was not an officially recognized title, but rather an informal lifetime pass from the government, giving him tacit permission to engage in sport in any protected forest he liked. And Jim Corbett, always the avid sportsman, certainly took advantage of it. During the 1930s and '40s, he was among the most influential men in the United Provinces, rubbing shoulders with lieutenant governors, maharajas, and movie stars alike, and whenever he could, he would take them hunting in the hills.

But all of these things came at a steep price. By all accounts, Corbett was an exceedingly humble and quiet man, never bragging about his exploits or explicitly using them for personal gain—in fact, even the proceeds from *Man-Eaters of Kumaon* were originally intended to benefit a charity for the blind. And with few exceptions, he personally eschewed the notion of hunting tigers for sport, refusing to shoot a tiger unless it could be certified as a serial man-eater. But regardless of his intentions, or the scores of human lives he actually saved, his reputation had been built directly upon the killing of tigers. And it was becoming increasingly clear to Corbett during the latter part of his life just how uncertain the future of the Indian tiger had become.

In 1907, when he had hunted the Champawat, there were probably close to 100,000 tigers left in the wild, more than half of which would have been found in India. By 1946, however, Corbett would confide in a letter to his friend Lord Wavell that in his opinion, fewer than four thousand wild Indian tigers remained. It was a slaughter that Jim Corbett had witnessed firsthand, arguably even contributed to, be it through his early work harvesting timber for the railroads, or as a government-contracted shikari organizing large-scale tiger hunts for visiting dignitaries. It was a revelation that may have come late for Corbett, but it possessed him fully.

In the last two decades of his life, Corbett devoted himself to the preservation of Indian wildlife—the Bengal tiger, in particular. The legendary tiger hunter became the animal's most dedicated conservationist. Using his connections with government officials, not to mention rare films of wild tigers that he himself had made, he actively lobbied the colonial British government to install protections for wild tiger populations, including his own brainchild, Hailey National Park, just outside of Nainital. He continued to give talks and make guest appearances at society functions, but increasingly he spoke as a naturalist bent on saving wild tigers and less as a "white hunter" who had spent much of his life dispatching them. His wildlife films and images, inspired by the early work of F. W. Champion, were pioneering feats of wildlife photography, and they served as tremendous aids in raising awareness of the plight of the Indian tiger. The man who had once carried a Martini-Henry rifle through the jungles of Kumaon now toted a 16-millimeter camera; rather than snarling tiger heads to mount on his wall, he now had pristine black-and-white photographs of the majestic cats. And Jim Corbett, who had loved and respected India's wildlife since childhood, came to realize that he much preferred the latter.

But India itself was changing. In the aftermath of the Second World War, after years of protests and petitions, the nation joined a number of other former colonies of the British Empire in achieving its independence. And what would become a moment of jubilation for much of the subcontinent proved to be a time of deep conflict and doubt for its colonial population—particularly those who, like the Corbetts, had been born and raised in India. In the uncertainty that followed, many gave in to their old fears, recalling the whispered tales they had grown up with of the bloody Rebellion of 1857. Fearing violent reprisals (never to materialize), or at the very least the same sort of discrimination that Indians had been subjected to for more than three centuries, most of those of European descent abandoned their adopted country and joined in a postcolonial exodus. Some returned to ancestral homes in the British Isles, others fled to the New World. And others, like Jim Corbett and his sister Maggie, sought out the remaining British colonies of East Africa.

The decision was anything but easy. Regardless of Corbett's background or family history, India was his home. And once again, he must have felt torn between two worlds. But with most of their British acquaintances fleeing in droves, and the future of an independent India uncertain, the Corbetts hastily sold off their properties and much of their belongings. They boarded a steamer bound for Mombasa in December 1947, leaving India behind them for good.

On June 5, 1950, a homesick Jim Corbett, on the cusp of turning seventy-five, wrote to his Kumaoni friend Jai Lal Sah, who was suffering from a serious illness back in Nainital. In his get-well note, dispatched from his new lodgings in Kenya, Corbett reflected:

You and I, and Hira Lal, are all that are left now of the old Kumaonis. Those were happy days for all of us and it is nice to think now that though we differed occasionally on municipal matters, we remained good friends, and as good friends we will meet again on the other side of the river where we will find Krishna Sah, Kundan Lal, Kishori Lal, Ram Singh, Mathura

Datt and many others who have gone before us but whom we still remember as good and valued friends.

The river he mentions is of course the Ganges—a river he had spent much of his adult life upon, but would never see again. And the old friends he calls by name are departed Hindus—fellow Indians—whom he hopes to meet again someday on its shores. Much is revealed in this short and touching passage from Corbett, an enigmatic man who even his own friends would admit was difficult to know.

That Jim Corbett felt compelled to leave India on the brink of its independence is a bittersweet ending, but in its own way, perhaps, a fitting one. He was simply unable to adapt to a changing world. In this regard, he was always more like the marginalized tigers he tracked through the forest than the well-adjusted human beings who crossed his path in the cities. A natural loner, a leftover from another era, he seemed to prefer a life of solitary exile on the world's last wild fringes rather than joining in society's industrialized march forward. Jim Corbett never married, having spent most of his prime either at isolated railway outposts or camping out in jungles, nor did he ever have any children. There have been numerous explanations for his lifelong bachelorhood posited over the years, and some of them may even hold water, although the most obvious is likely the truest of all: the man simply preferred to be alone. In later years, at his small estate in Kaladhungi, he would insist on sleeping by himself in a tent beside his brand-new house, feeling more at ease on the edge of the jungle than safe inside any man-made walls. He had been born an outsider, and to a certain extent, despite his many awards and decorations, he would die one as well.

When his heart gave out on April 19, 1955, in Nyeri, Kenya, the old Kumoani had one final honor awaiting him. The following year, Hailey National Park, the protected wildlife sanctuary he had helped establish outside of Nainital, was renamed Jim Corbett National Park in his honor—an extraordinary show of appreciation from an independent Indian government that still held the Carpet Sahib, as he was known to his Kumaoni friends, in high esteem. And if there is an honest testament to the quality of the man, it is almost certainly this: that while in the end he may never have felt completely a part of either British or Indian society, he would finish his life beloved by both.

As for the persistence of wild tigers in India, they too are a living testament of sorts. Jim Corbett was among the first to call attention to the plight of the tiger, at a time when many still saw them as nothing more than a menace to be exterminated. His early efforts laid the groundwork for the establishment of India's first protected tiger reserves, as well as the Project Tiger conservation initiative launched by the Indian government in 1973. His writings and photographs as a naturalist inspired a whole generation of tiger conservationists and provided seminal insights into the stresses environmentally isolated tigers can face—insights that today have led to actions such as the Terai Arc Landscape (TAL) project in Nepal, which is working in collaboration with local populations to re-wild corridors between tiger sanctuaries across the *terai*, in an effort to create a contiguous and sustainable tiger habitat. It is a daunting project, but one that, if successful, promises to secure the species a long-term future, at least in the region.

As of the writing of this book, estimates put the population of wild tigers across all of Asia at close to four thousand. It is a precarious number, and the species is still gravely threatened. A growing luxury market in China with a demand for tiger skins and other body parts, further aggravated by the highly controversial practice of captive tiger farming, has taken its toll in neighboring countries, encouraging poaching in Russia, India, and Nepal—in fact, tiger poaching in Indian forests for 2016 surged to its highest level in fifteen years. And in places where poaching is not a primary culprit, simple habitat destruction proves just as deadly. Since 2010, encroaching farmland has caused tigers to all but vanish in Laos, Vietnam, and Cambodia, and the rise of palm oil plantations in Sumatra has pushed the subspecies of tigers there to the brink of extinction—something that has already happened in the nearby islands of Java, Bali, and Singapore. According to tiger conservationist Ullas Karanth, apart from a few reserves in India and Thailand, "there are no convincing data to show that populations are recovering in the rest of Southeast Asia or Russia." With the wild tiger's range having collapsed by 93 percent of what it once was, and with their total number only a small fraction of what it has been in centuries past, the animal's future is imperiled, to say the least.

But there is some good news. Despite the dangers that wild tigers still face, particularly smaller, isolated populations in East Asia, for the first time in more than a century, their total numbers appear to be growing. According to a recent census lauded by the World Wildlife Fund, the global population increased from an estimated 3,200 wild tigers in 2010, to 3,890 wild tigers in 2016, with Jim Corbett's beloved Indian Bengals leading the charge, making up 70 percent of the aggregate total. There has been some debate as to how accurate the numbers actually are, and considerable concern on the part of some tiger experts who feel that the report paints an overly cheerful picture, even labeling it a "disservice to conservation." But one aspect even critics of the survey seem to agree on is that in large, well-protected wildlife reserves, in India in particular, local tiger populations are reasonably stable—in some cases even growing—and serve as the best hope for the future of the species. And in the reserves of the Malenadu region, one recent study even suggested that the local tiger population had increased fivefold in the last fifty years, likely making it the largest in the world—a feat made possible by the dedicated conservation efforts of the government and surrounding communities. And even if smaller subspecies populations are under threat elsewhere in Asia, the governments of the countries that harbor them are at the very least aware of the problem and attempting to take countermeasures. At a 2010 summit, a number of these governments came together to pledge their common goal of doubling wild tiger numbers by 2022—a daunting challenge, but one that's not totally beyond the realm of possibility. And simply stabilizing the endangered sub-populations would be a landmark event. As to whether or not that happens remains to be seen.

The role of apex predators in a healthy and diverse ecosystem is one that these days is seldom debated. We've arrived at the point, it seems, where all can agree that tigers need to exist. The challenge remaining, however, is managing our existence alongside theirs, even in places where tiger populations are relatively secure. Human–tiger conflict continues to pose problems in areas in India and Nepal where the two species overlap, and incentivizing the preservation of tigers in communities that have to live with them every day is a persistent challenge. Yet there is a potential symbiosis of sorts, one that many Tharu and Pahari peoples seem to comprehend intuitively. When the forest is seen not as an obstruction to be cleared, but rather as a crucial resource to be preserved, a convincing incentive becomes apparent. And when it comes to guarding and maintaining the health of a forest, there is no better partner than a wild tiger. Jim Corbett was unique for his time and place in that he understood this all too well. Despite seeing the bodies of countless victims, despite nearly becoming one himself on multiple occasions, he never once harbored ill will toward the predatory cats—if anything, he exhibited profound empathy for their plight. He understood the secret truth about humans and tigers. It is the same truth that was chiseled on Sanskrit tablets all those millennia ago, and even hinted at by William Shakespeare in the epigraph of this book:

That when it comes to truly behaving as a beast—to killing wantonly and without reason—it is our kind, not theirs, that is the fiercer of the two.

EPILOGUE

||||||||||||||||||||

Given the extraordinary nature of the Champawat Tiger, and its even more extraordinary alleged human tally, it seemed worthy to include a brief coda for those interested in the available solid documentation surrounding the tiger. As Jim Corbett himself admitted, the total of 436 human victims was an informed estimate based on reported attacks that occurred in its home range, during the time it was assumed to be active. It almost certainly is not a precise number, which is to be expected when it comes to a jungle predator that hunted on a regular basis for nearly a decade, in one of the more remote corners of the Himalayan foothills over a century ago. In actuality, the sum could have been somewhat lower, or even, given the spotty nature of written records at that time, *significantly* higher. Still, there is ample evidence worth considering when trying to deem the alleged total as being at the very least credible, if not fully verified by hard documentary sources. Granted, the Nepalese portion of the tiger's spree will likely always be hazy; the boyhood recollections of the elderly Nara Bahadur Bisht regarding the Rupal Man-Eater at the turn of the twentieth century certainly do support the story, even down to Corbett's claim that it was chased out of Nepal by armed men. And indeed, oral traditions such as this are crucial in trying to understand the tiger's origins—even today, stories are told in the region, passed down from generation to generation, of the legendary man-eater that once terrorized the population. However, locating printed documentation of Nepalese

tiger attacks from that period is next to impossible. Written records of any kind from Rana-era western Nepal are rare, and those
that concern tiger attacks, virtually unheard of. In colonial India,
though, where the eradication of tigers and human–tiger conflict
had both a practical and symbolic importance to government officials, more detailed records were kept—and in the copious stacks
of the British Library, some can even be found today.

That there was an exceptionally dangerous man-eating tiger in
Kumaon, and that it was finally killed by Jim Corbett in 1907,
the historical record leaves little room for doubt. The most explicit
and detailed evidence I was able to find to corroborate Corbett's
account from *Man-Eaters of Kumaon* came in the form of an article
that appeared in the Dehradun edition of the Indian periodical
The Pioneer, published on June 7, 1907. The article makes mention
of the tiger that was killed in April by Wildblood and that was mistakenly believed to be the Champawat, but goes on to state quite
clearly that it was not "till the 12th of May that Mr. Corbett of
Naini Tal succeeded in shooting it." The story upholds Corbett's
claim that other hunters had tried to kill the tiger without success,
that he was assisted by the Tahsildar as well as the Tahsildar's assistant patwari, who are both identified by name, and it gives an
account, very similar to Corbett's, of the tiger's final victim and the
beat that was organized to flush it out of the ravine. The fact that
the tiger had lost "both the upper and lower canine teeth on one
side" is mentioned as well.

There is also journalistic evidence of the tiger's attacks prior
to Corbett's final hunt—including those attacks mistakenly attributed to the tiger that Wildblood killed just before Jim Corbett's involvement. In an article from *The Times of India,* dated
April 15, 1907—almost a full month before Corbett killed the

Champawat—mention is made of a man-eating tiger that had been killed near Lohaghat during the "cold weather" by an "English shikari" who would have been Wildblood. The article blames man-eating tiger(s) for having "kept the villages scattered over a wide range of country in a state of continual terror." It goes on to admit that the tiger Wildblood killed, although possibly a man-eater as well, was not the creature being sought, as "the depredations have not ceased" and "another tiger of the same kind is at large." The article emphasizes the unusually high number of victims attributed to the tiger, and the fact that it was published in Bombay (today Mumbai) shows that the news of the Champawat's work was clamant enough to have spread well beyond the hills of Kumaon, all the way to the press clubs of major Indian capitals.

In addition to the periodicals of the day, there is also evidence to be found in official government records specifically regarding the engraved rifle that was given to Corbett as a reward for killing the Champawat. A British report titled "United Provinces, Agra and Oudh Proceedings" dated 1907 shows an expenditure of "Rs. 463·6·0" specifically for the "purchase of arms for presentation to certain persons for killing a man-eating tiger at [Champawat]." This would have included the ornamental knife given to the patwari who accompanied Corbett during his initial pursuit of the tiger, the engraved gun presented to the Tahsildar for his brave service during the final beat, and the .275 Rigby rifle that was presented to Jim Corbett by "Sir J. P. Hewett," who was lieutenant governor of the United Provinces at that time. The Rigby rifle still exists, and it is in the possession of the gunsmiths at John Rigby & Company in London, where it is frequently put on display. The engraved silver plaque on its stock states that it was given to Jim Corbett "in recognition of his having killed a man-eating tigress at Champawat in 1907." The United Provinces government gazette from the same year makes mention of the engraved arms as "special rewards" that were offered "for the destruction of a known man-eating tiger" in Champawat; additionally, the gazette's report on the destruction of wild animals and venomous snakes from 1907 records an offering of "special rewards in the shape of firearms" that were given out by the lieutenant governor for dispatching a man-eater in Champawat's vicinity. It warrants mention that an engraved sporting rifle of this quality, presented to a domiciled railroad employee like Corbett by a lieutenant governor, was a highly unusual gift—its very existence is evidence not only of Corbett having finally killed the tiger, but also of this particular tiger's unique notoriety. That an engraved gun was also gifted to an Indian Tahsildar from a historically rebellious region is also unusual, perhaps even more so. High-ranking government officials rarely concerned themselves with specific man-eaters, let alone the lowly shikaris who dispatched them, so it stands to reason that this man-eater in particular was exceptionally lethal, a tiger unlike any they had ever dealt with before.

As far as the actual number of victims the Champawat claimed in Kumaon, while there is no doubt that it was sizable, there is some variance as to just how sizable it actually was. As previously noted, Corbett claimed that the tiger was responsible for 436 human kills, 236 of which he attributed to its time in Kumaon. The Nepalese portion is essentially impossible to verify with documentary evidence, as virtually none exists. There are, however, Indian records which do indeed show that the tiger was unusually prolific, although each account seems to provide a slightly different number of victims. In the aforementioned *Times of India* article from April 15, 1907, the claim is made that "about a hundred women are known to have been killed and that the total number is probably much higher." This account, attributing at *least* a hundred victims to the Champawat in Kumaon, and likely many more, seems to lend some credence to Corbett's account of the Indian portion of the Champawat's spree. However, the version provided by *The Pioneer* article from June 7, 1907, gives a slightly lower figure, attributing the loss of "about 70 persons, nearly all females" to the tiger, or possibly even multiple tigers, as the idea of a single animal causing so much carnage was evidently hard to believe.

As for the reason for the divergence, there are all manner of possibilities—and the possibility that Corbett was mistaken (or even exaggerating) should at least be considered, although it seems he would have been in a much better position to get honest and accurate firsthand data from the field than would most big-city reporters, and he definitely wasn't known as a teller of tall tales. One plausible explanation for the variance in numbers could simply be the source of the journalist's data—specifically, if it was coming from government sources, including the official administrative gazettes. Evidence of the Champawat Tiger does indeed appear in the colonial government's documents regarding deaths from wild animals and venomous snakes in Kumaon. However, such records are notorious for their inaccuracy when it comes to the true number of attacks, for a simple reason: a very large number of tiger attacks, especially those that occurred in remote mountain regions where anti-British sentiment was strong, went unreported to the government. And even attacks that were reported and were *strongly* suspected to be the work of a tiger were often dismissed by authorities if an eyewitness or bodily remains could not be found. With tangible verification so difficult to procure, it hardly seems likely that a deputy commissioner interested in preserving his reputation would go out of his way to prove that a tiger attack occurred and risk looking incompetent. After all, given the symbolic associations unruly tigers had to the colonial mind, few government officials would have had any interest in advertising their inability to control their "tiger problem." Accordingly, the underreporting of fatal tiger attacks was rampant—the government numbers are useful, as they are often the only records available, but they should always be taken with a healthy grain of salt.

With that in mind, what follows are the annual recorded tiger fatalities in the Kumaon division, as reported by the United Provinces government gazettes from each year:

1900: 3 victims
1901: 3 victims
1902: 6 victims
1903: 6 victims
1904: 4 victims
1905: 4 victims
1906: 20 victims
1907: 39 victims

Given the tremendous leap in the number of victims, these figures make the Champawat's presence all but undeniable. Clearly, something out of the ordinary was happening in the hills of Kumaon. Worth noting, however, is that while Corbett claimed he first heard mention of the tiger four years prior to killing it, the government records only show a sharp increase in fatal attacks the final *two* years, providing a total tally and time line that seem more on par with what the *The Pioneer,* and to a lesser extent *The Times of India,* reported. Does this mean that Corbett was simply mistaken? Or could it also be that the colonial government was simply unaware of—or even initially reluctant to report on—a large number of fatal tiger attacks that occurred in a remote and historically hostile region of the Empire? Indeed, is it even possible that the *actual* numbers Jim Corbett heard whispered in confidence from government officials like Deputy Commissioner Charles Henry Berthoud were drastically different from the more palatable totals that the colonial government allowed the newspapers to print and the public to see?

It's difficult to say, although historians and conservationists alike have cited Jim Corbett's estimate of 436 total deaths in Nepal and India as credible over the years, and the hunter-turned-naturalist was always respected for his honesty when it came to his reporting on tigers. The fact that scientific studies of tiger populations still cite many of his findings close to a hundred years on bears witness to that fact. Some of his later man-eater hunts from the 1920s and '30s are much better documented, going so far as to catalog tiger attacks by village and date. The hunt for the Champawat Tiger, however, occurred very early in his career, and when writing about the encounter nearly four decades later (*Man-Eaters of Kumaon* wasn't published until 1944), such details were simply unavailable. Modern-day man-eaters, such as the tiger that stalked Baitadi, Nepal, in 1997, and the tiger that escaped India's Jim Corbett National Park in 2014, have clearly shown that man-eating on a large scale, over extended periods of time, is certainly possible in the region, even in the present day. If a man-eater could still cause one hundred human fatalities at the dawn of the twenty-first century, it only makes sense that in the early 1900s, in an era without electric lights, telecommunications, or motorized transportation, such a tiger would have been that much more lethal, and that much harder to find and kill.

Admittedly, ironclad proof of the Champawat's egregious number will likely remain elusive. It will probably always be an alleged number, albeit one provided by one of the few English-speaking individuals who lived in the Kumaoni hills, understood the local languages, had a vast knowledge of tigers, and had direct access to high-ranking officials and indigenous community leaders alike—essentially, the most credible source on man-eaters available from that time and place. Additionally, it was a number that Lord Linlithgow, the Viceroy of India from 1936 until 1943, was comfortable attaching his name to—he vouched for the accuracy of Corbett's accounts in the preface he wrote for *Man-Eaters of Kumaon*. Nevertheless, it is still a number that ought to be scrutinized and authenticated to whatever extent the historical record allows, and new archival sources could readily change the story. There is always the possibility that an old government report or newspaper article could turn up that adds more clarity to a narrative that time and distance have conspired to obscure. Just in researching this book, details were discovered in the British Library's ample store of colonial records regarding the hunt for the Champawat that had previously been lost to the ages. So the possibility always exists.

There is, however, one other potential way of authenticating the number that at least deserves mention—one that ultimately relies less on dusty libraries than sterile laboratories, and involves the location of the tiger itself. While conducting research in India, I visited the famous Gurney House, which once served as the Nainital residence of the Corbett family. The current owners keep some of Jim Corbett's old possessions and assorted memorabilia on display to preserve the house's history. Among the many framed pictures, there is a black-and-white photograph of what appears to be the Champawat Tiger's taxidermied head, damaged teeth and all—evidence that the preserved skin was indeed brought back and kept by Corbett. The same photo also appears in the second printing of a book published in India in 1999 titled *Jim Corbett of Kumaon*. It was authored by D. C. Kala, whose own father had known the famous hunter personally, and it clearly identifies the tiger head and skin in question as belonging to the Champawat Man-Eater. And if that weren't enough, the photograph's authenticity appears to be verified by a second photo, taken in 1926 and currently in the possession of the Ibbotson family (Sir William "Ibby" Ibbotson was a close friend of Jim Corbett), which shows the same stuffed head from another angle, alongside the skins of other man-eaters outside of Corbett's home. Locating the *current* whereabouts of the Champawat skin, however, has proven more difficult than authenticating the photograph. Most of Corbett's trophies were either given away as gifts, or sold at auction in Africa shortly after his death, with the proceeds going to various charities in Kenya and Kumaon. The location of only a few of them are known today, and many have been lost or destroyed over the years. But if the head and skin of the Champawat do still exist, and if they were to be located, it is theoretically possible that a laboratory test could provide an answer as to how many human victims the tiger actually ingested, based on the chemical signatures of its bones and hair. It was precisely this sort of scientific analysis that was used to determine the number of railway workers that the infamous Lions of Tsavo consumed in 1898, and it's entirely possible that a similar test performed on the preserved head of the Champawat Tiger could provide insight into its own human tally. Until that day, however, it is probable that the estimate of its total kills will remain just that. Credible, perhaps even likely, but impossible to prove beyond a shadow of a doubt.

But that the Champawat Tiger existed, hunted and ate human beings by the dozens, and was eventually shot by Jim Corbett in 1907 is an established fact. The number may beg further study, but the story is true. And at the bottom of the Champa Gorge, near the village of Phungar just outside of Champawat proper, the projecting rock where Corbett killed the man-eater still stands today— just as he described it—a mute witness to the dramatic events that transpired there well over a century ago.

ACKNOWLEDGMENTS

||||||||||||||||||||

Uncovering and re-creating the exploits of a man-eating tiger more than a century later is a challenging endeavor, and this book only exists thanks to the encouragement and support of some exceedingly generous individuals. For their archival work, particularly in regards to the India Office Records of the British Library, I would like to thank Justine Taylor and MacKenzie Gibson. For their assistance in Nepal, I owe a tremendous debt to Sanjaya Mahato, Dr. Gisèle Krauskopff, the *guraus* Sukh Lal Chaudhary and Kanan Chaudhary, the *patriti* Hwae, and the *pujari* Bashanta. In India, the expertise of Kamal Bisht and the hospitality of the Dalmia family was crucial for researching historic sites, and many thanks are owed to them as well. In parsing out the details of Jim Corbett's life, the counsel of Jerry Jaleel, Peter Byrne, Dr. Joseph Jordania, and Kotetcha Kristoff proved extremely helpful. And in terms of actually transforming all of this into a published work, I could not have done so without the backing of my agent, Jim Fitzgerald; my editor, Peter Hubbard, who suggested the Champawat Tiger as a subject; and the rest of the team at William Morrow/HarperCollins, including Liate Stehlik, Nick Amphlett, Lauren Janiec, and Ryan Cury. Last, but certainly not least, I would like to offer a warm and heartfelt *dhanyavaad* to the many Tharu and Kumaoni people who were kind enough to share their lives, their homes, and their stories with a stranger. A portion of the proceeds of this book will be donated to Chitwan's Tharu Cultural Museum & Research Center, as well as the local Tharu Wildlife Initiative—both of which are committed to preserving the region's natural heritage, including its tigers.

BIBLIOGRAPHY

||||||||||||||||||

CHAPTER 1:

Quammen, David. *Monster of God*. New York: Norton & Company, 2003.

Mishra, Hemantha. *Bones of the Tiger*. Guilford: Lyons Press, 2010.

Nikolaev, Igor and Victor Yudin. "Conflicts Between Man and Tiger in the Russian Far East." *Bulletin Moskovskogo obshchestva ispytateley Prirody,* vol. 98, issue 3, 1993.

Seidensticker, John. *Riding the Tiger: Tiger Conservation in Human-Dominated Landscapes*. Cambridge: Cambridge University Press, 1999.

Conover, Adele. "The Object at Hand." *Smithsonian,* November 1995.

Goldman, Adam. "New respect for tiger leaping ability." *Los Angeles Times,* January 13, 2008.

Koopman, John. "S.F. cops tell how they killed raging zoo tiger." *San Francisco Chronicle,* February 4, 2009.

Naumov, N. P. and V. G. Heptner. *Mammals of the Soviet Union*. Moscow: Vysshaya Shkola Publishers, 1972.

"Rare incident: amur tiger reportedly hunting for seals." *Pravda,* December 20, 2002.

"Tiger kills adult rhino in Dudhwa Tiger Reserve." *The Hindu,* January 29, 2013.

"Kaziranga elephant killed in tiger attack." *The Times of India,* October 22, 2014.

Lenin, Janaki. "Hunting by mimicry." *The Hindu,* August 9, 2013.

C H A P T E R   2 :

Locke, Piers. "Food, ritual, and interspecies intimacy in the Chitwan elephant stables." *The South Asianist,* vol. 2, no. 2, 2013.

Krauskopff, Gisèle and Pamela Meyer. *The Kings of Nepal and the Tharu of the Tarai: Fascimiles of Royal Land Grant Documents issued from 1726–1971 AD.* Los Angeles: Rusca Press, 2000.

Guneratne, Arjun. "The Shaman and the Priest: Ghosts, Death, and Ritual Specialists in Tharu Society." *Himalaya,* vol. 19, no. 2, 1999.

Chowdhury, Arabinda, Arabinda Brahma, Ranajit Mondal, and Mrinal Biswas. "Stigma of tiger attacks: Study of tiger-widows from Sunderban Delta, India." *Indian Journal of Psychiatry,* 58:12–9, 2016.

Raffaele, Paul. "Man-Eaters of Tsavo." *Smithsonian,* January 2010.

"Tiger Kills 35 Children in Western Nepal District." Other News to Note: The World. *Orlando Sentinel,* January 25, 1997.

"Nepal Officer Sentences Man-Eating Tiger to Death." Other News to Note: The World. *Orlando Sentinel,* July 1, 1997.

"Science File: Nepalese Man-Eater." *Los Angeles Times,* November 13, 1997.

Kumar, Hari and Ellen Barry. "Tiger Population Grows in India, as Does Fear After Attack." *New York Times,* February 11, 2014.

Pfalz, Jennifer. "Man-Eating Tigers on the Prowl in India." *Liberty Voice,* February 12, 2014.

Banerjee, Biswajeet. "Man-Eating Tiger Claims 10th Victim in India." *The Star,* February 10, 2014.

Gurung, Bahadur Bhim. "Ecological and Sociological Aspects of Human-Tiger Conflicts in Chitwan National Park." Submitted to the faculty of the graduate school of the University of Minnesota, July 2008.

Zielinski, Sarah. "Secrets of a Lion's Roar." *Smithsonian,* November 3, 2011.

Pathak, Hrishikesh, Jaydeo Borkar, Pradeep Dixit, Shaildendra Dhawane, Manish Shrigiriwar, and Niraj Dingre. "Fatal tiger attack: A case report on emphasis on typical tiger injuries characterized by partially resembling stab-like wounds." *Forensic Science International*. Published by El-sevier. First available online August 16, 2003.

Sunquist, Fiona and Mel Sunquist. *Tiger Moon*. Chicago: University of Chicago Press, 1988.

Vaillant, John. *The Tiger: A True Story of Vengeance and Survival*. New York: Vintage Books, 2010.

Mishra, Hemantha. *Bones of the Tiger*. Guilford: Lyons Press, 2010.

Chakraborty, Monotosh. "Tiger snatches man off boat, leaps back into Sundarbans jungle." *The Times of India*, June 27, 2014.

Thapar, Valmik. *Tiger: Portrait of a Predator*. Surrey: Bramley Books, 1986.

Beveridge, Candida. "Face to face with a man-eating tiger." *BBC Magazine*, November 12, 2014.

CHAPTER 3:

Guneratne, Arjun. *Many Tongues, One People: The Making of Tharu Identity in Nepal*. Ithaca: Cornell University Press, 2002.

Nara Bahadur BishtBorah, Jimmy. "Tigers of the Transboundary Terai Arc Landscape: Status, Distribution, and Movement in the Terai of India and Nepal." New Delhi: Global Tiger Forum, 2014.

Guneratne, Arjun. "The Shaman and the Priest: Ghosts, Death and Ritual Specialists in Tharu Society." *Himalaya, the Journal of the Association for Nepal and Himalayan Studies,* vol. 19, no. 2, 1999.

Locke, Piers. "The Tharu, the Tarai and the History of the Nepali Hattisar." *European Bulletin of Himalayan Research,* 38:59–80, 2011.

Bell, Thomas. "Diary of a disastrous campaign." *HIMAL,* December 21, 2012.

Krauskopff, Gisèle and Pamela Meyer. *The Kings of Nepal and the Tharu of the Tarai: Fascimiles of Royal Land Grant Documents issued from 1726–1971 AD*. Los Angeles: Rusca Press, 2000.

Adhikari, Krishna Kant. "A Brief Survey of Nepal's Trade with British India

During the Latter Half of the Nineteenth Century." *INAS Journal*, vol. 2, no. 1, February 1975.

Locke, Piers. "The Tharu, the Terai and the History of the Nepali Hattisar." *European Bulletin of Himalayan Research*, 38:59–80, 2011.

Schaller, George. *The Deer and the Tiger*. Toronto: University of Chicago Press, 1967.

Mishra, Hemantha. *Bones of the Tiger*. Guilford: Lyons Press, 2010.

Thapar, Valmik. *Tiger: Portrait of a Predator*. Surrey: Bramley Books, 1986.

Corbett, Jim. *Man-Eaters of Kumaon*. New Delhi: Oxford University Press, 1988.

Byrne, Peter. *Gentleman Hunter*. Long Beach: Safari Press, 2007.

CHAPTER 4:

Corbett, Jim. *Man-Eaters of Kumaon*. New Delhi: Oxford University Press, 1988.

Segrave, Bob, ed. *Behind Jim Corbett's Stories: An Analytical Journey to "Corbett's Places" and Unanswered Questions*. Tbilisi: Logos, 2016.

Thapar, Valmik. *Land of the Tiger*. Berkeley: University of California Press, 1997.

Rangarajan, Mahesh. *India's Wildlife History: An Introduction*. Delhi: Permanent Black, 2001.

Mitra, Sudipta. *Gir Forest and the Saga of the Asiatic Lion*. New Delhi: Indus Publishing Company, 2005.

Judd, Denis. *The Lion and the Tiger: The Rise and Fall of the British Raj, 1600 to 1947*. New York: Oxford University Press, 2010.

Forbes, James. *Oriental Memoirs: A Narrative of Seventeen Years Residence in India*. London: Published by Richard Bentley, 1834.

Narrative Sketches of the Conquest of Mysore Effected by the British Troops and Their Allies. Printed by W. Justins, May 4, 1799.

Sunquist, Fiona and Mel Sunquist. *Tiger Moon*. Chicago: University of Chicago Press, 1988.

Imperial Gazetteer of India, vol. 1, p. 218. Oxford: Clarendon Press, 1909. Published under the authority of his majesty's secretary of state for India in council. Digital South Asia Library, University of Chicago.

Williamson, Thomas. *Oriental Field Sports*, vol. 1. London: Printed by Edward Orme, 1807.

Mishra, Hemantha. *Bones of the Tiger*. Guilford: Lyons Press, 2010.

Vaillant, John. *The Tiger: A True Story of Vengeance and Survival*. New York: Vintage Books, 2010.

"Our Colonies, No. V: The Timbers of our Indian Possessions and Australia." *The Nautical Magazine for 1875*, Vol. XLIV. London: Simpkin, Marshal & Co., 1875.

Mittal, Arun K. *British Administration in Kumaon Himalayas: A Historical Study 1815–1947*. Delhi: Mittal Publications, 1986.

Spinage, C. A. *Cattle Plague: A History*. New York: Kluwer Academic/Plenum Publishers, 2003.

"Reports on the destruction of wild animals & venomous snakes during the year 1908." IOR/L/PJ/6/971, File 4068, Letter No. 74 from India. Judicial and Public Department, British Library.

"Reports on the destruction of wild animals and venomous snakes during the year 1907." IOR/L/PJ/6/893, File 3661. India Office Records and Private Papers, British Library.

"Reports on the destruction of and mortality caused by wild animals & snakes during 1906." IOR/L/PJ/6/6830, File 3349, Letter No. 73. Judicial and Public Department, British Library.

"Reports on the destruction of, and mortality caused by wild animals and snakes during 1906." IOR/L/PJ/6/830, File 3349. India Office Records and Private Papers, British Library.

"A Wronged Animal: Justice for the Tiger." *The Times of India,* p. 10, October 13, 1908.

LaFreniere, Gilbert. *The Decline of Nature: Environmental History and the Western Worldview.* Palo Alto: Academic Press, 2008.

Ogilvy, D. *A Book of Highland Minstrelsy.* London: Richard Griffin and Company, 1860.

CHAPTER 5:

Corbett, Jim. *Man-Eaters of Kumaon.* New Delhi: Oxford University Press, 1988.

"Man-Eating Tigers." *The Times of India,* p. 8, April 15, 1907.

"Man-Eaters in Kumaon." *The Pioneer,* June 7, 1907.

Segrave, Bob, ed. *Behind Jim Corbett's Stories: An Analytical Journey to "Corbett's Places" and Unanswered Questions.* Tbilisi: Logos, 2016.

Eardley-Wilmot, Sainthill. *Forest Life and Sport in India.* London: Edward Arnold, Publisher to H. M. India Office, 1910.

Sramek, Joseph. "Face Him Like a Briton: Tiger Hunting, Imperialism, and British Masculinity in Colonial India, 1800–1875." *Victorian Studies,* vol. 48, no. 4, Summer 2006. Indiana University Press.

Storey, William. "Lion and Tiger Hunting in Kenya and Northern India, 1898–1930." *Journal of World History,* vol. 2, no. 2, Fall 1991. University of Hawaii Press.

CHAPTER 6:

Corbett, Jim. *Man-Eaters of Kumaon.* New Delhi: Oxford University Press, 1988.

Rawat, Ajay. *Forest Management in Kumaon Himalaya: Struggle of the Marginalized People.* New Delhi: Indus Publishing Company, 1999.

"Man-Eating Tigers." *The Times of India,* p. 8, April 15, 1907.

Sunquist, Mel and Fiona Sunquist. *Wild Cats of the World.* Chicago: University of Chicago Press, 2002.

"Reports on the destruction of wild animals & venomous snakes during the year 1908." IOR/L/PJ/6/971, File 4068, Letter No. 74 from India. Judicial and Public Department, British Library.

"Reports on the destruction of wild animals and venomous snakes during the year 1907." IOR/L/PJ/6/893, File 3661. India Office Records and Private Papers, British Library.

Sharma, Sandeep, Yadvendradev Jhala, Yadvendradev, and Vishwas Sawarkar. "Gender Discrimination of Tigers by Using their Pugmarks." *Wildlife Society Bulletin,* vol. 31, no. 1, Spring 2003.

CHAPTER 7:

Corbett, Jim. *The Second Jim Corbett Omnibus.* "Jungle Lore." New Delhi: Oxford University Press, 1991.

CHAPTER 8:

Booth, Martin. *Carpet Sahib: The Life of Jim Corbett.* London: Constable Press, 1986.

Kala, D. C. *Jim Corbett of Kumaon.* Mumbai: Penguin Books, 2009.

James, Lawrence. *Raj: The Making and Unmaking of British India.* London: Little, Brown and Company, 1997.

Judd, Denis. *The Lion and the Tiger: The Rise and Fall of the British Raj, 1600 to 1947.* Oxford: Oxford University Press, 2010.

"Dangers of the Jungle: The Human Death Toll." *The Times of India,* p. 9, September 29, 1906.

"Natives and Fire-Arms." *The Times of India,* p. 8, April 15, 1907.

"Man-Eaters in Kumaon." *The Pioneer,* June 7, 1907.

Kipling, Rudyard. *The Phantom Rickshaw & Other Eerie Tales.* London: A. H. Wheeler & Co., 1888.

CHAPTER 9:

Schaller, George. *The Deer and the Tiger*. Toronto: University of Chicago Press, 1967.

CHAPTER 10:

Corbett, Jim. *The Second Jim Corbett Omnibus*. "Jungle Lore." New Delhi: Oxford University Press, 1991.

Booth, Martin. *Carpet Sahib: The Life of Jim Corbett*. London: Constable Press, 1986.

Kala, D. C. *Jim Corbett of Kumaon*. Mumbai: Penguin Books, 2009.

"Reports on the destruction of, and mortality caused by wild animals and snakes during 1906." IOR/L/PJ/6/830, File 3349. India Office Records and Private Papers, British Library.

"Man-Eaters in Kumaon." *The Pioneer,* June 7, 1907.

Corbett, Jim. *Man-Eaters of Kumaon*. New Delhi: Oxford University Press, 1988.

Petzal, David. "Black Powder Behemoths." *Field and Stream,* December 1, 2004.

CHAPTER 11:

Corbett, Jim. *Man-Eaters of Kumaon*. New Delhi: Oxford University Press, 1988.

Nayar, Pramod. *English Writing and India, 1600–1920: Colonizing aesthetics*. New York: Routledge, 2008.

Inglis, James. *Tent Life in Tigerland*. London: Low, Marston, and Company, 1892.

CHAPTER 12:

Corbett, Jim. *Man-Eaters of Kumaon*. New Delhi: Oxford University Press, 1988.

Kala, D. C. *Jim Corbett of Kumaon*. Mumbai: Penguin Books, 2009.

CHAPTER 13:

Kala, D. C. *Jim Corbett of Kumaon*. Mumbai: Penguin Books, 2009.

Guynup, Sharon. "As Asian Luxury Market Grows, a Surge in Tiger Killings in India." *Yale Environment 360,* January 10, 2017.

Leahy, Stephen. "Extremely Endangered Tiger Losing Habitat—and Fast." *National Geographic,* December 10, 2017.

Hauser, Christine. "Number of Tigers in the Wild is Rising, Wildlife Groups Say." *New York Times,* April 11, 2016.

Guynup, Sharon. "How Many Tigers Are There Really? A Conservation Mystery." *National Geographic,* April 20, 2016.

Karanth, Ullas K. "The Trouble with Tiger Numbers." *Scientific American,* July 1, 2016.

EPILOGUE:

"Man-Eaters in Kumaon." *The Pioneer,* June 7, 1907.

"United Provinces, Agra and Oudh Proceedings." 1907. IOR/P/7536. India Office Records and Private Papers, British Library.

"Deaths caused by wild animals and venomous snakes during the period 1900–1901 and measures adopted for their destruction." IOR/L/PJ /6/582, File 1967. India Office Records and Private Papers, British Library.

"Statistics for the destruction of wild animals and snakes during 1901." IOR/L/PJ/6/615, File 2169. India Office Records and Private Papers, British Library.

"Report on the destruction of wild animals and venomous snakes during 1902." IOR/L/PJ/6/648, File 2106. India Office Records and Private Papers, British Library.

"Report on the destruction of wild animals and snakes during 1903." IOR/L/PJ/6/701, File 2845. India Office Records and Private Papers, British Library.

"Reports on deaths caused by wild animals and venomous snakes during 1904." IOR/L/PJ/6/736, File 3177. India Office Records and Private Papers, British Library.

"Reports on the destruction of wild animals and snakes during the year 1905." IOR/L/PJ/6/781, File 3439. India Office Records and Private Papers, British Library.

"Reports on the destruction of, and mortality caused by wild animals and snakes during 1906." IOR/L/PJ/6/830, File 3349. India Office Records and Private Papers, British Library.

"Reports on the destruction of wild animals and venomous snakes during the year 1907." IOR/L/PJ/6/893, File 3661. India Office Records and Private Papers, British Library.

"Man-Eating Tigers." *The Times of India,* p. 8, April 15, 1907.

Segrave, Bob, ed. *Behind Jim Corbett's Stories: An Analytical Journey to "Corbett's Places" and Unanswered Questions.* Tbilisi: Logos, 2016.

INDEX

||||||||||||||||||||||

DANE HUCKELBRIDGE has written for the *Wall Street Journal*, *Tin House*, *The New Republic*, and *New Delta Review*. He is the author of *Bourbon: A History of the American Spirit*; *The United States of Beer: The True Tale of How Beer Conquered America, From B.C. to Budweiser and Beyond*; and a novel, *Castle of Water*, which has been optioned for film. A graduate of Princeton University, he lives in Paris.

More from Dane Huckelbridge

A rollicking biography of bourbon whiskey that doubles as "a complex and entertaining" (*WALL STREET JOURNAL*) history of America itself.

Few products are so completely or intimately steeped in the American story as bourbon whiskey. As Dane Huckelbridge's masterfully crafted history reveals, the iconic amber spirit is the American experience, distilled, aged, and sealed in a bottle.

Dane Huckelbridge's cultural history charts the wild, engrossing, and surprisingly complex story of our favorite alcoholic drink, showing how America has been under the influence of beer at almost every stage.

Drawing upon a wealth of little-known historical sources, explaining the scientific breakthroughs that have shaped beer's evolution, and mixing in more than a splash of dedicated on-the-ground research, *The United States of Beer* offers a raucous and enlightening toast to the all-American drink.

The astonishing true story of the tiger that claimed a record 437 human lives.

One part pulse-pounding thriller, one part soulful natural history of the endangered Royal Bengal tiger, *No Beast So Fierce* is Dane Huckelbridge's gripping nonfiction account of the Champawat tiger, which terrified northern India and Nepal from 1900 to 1907, and Jim Corbett, the legendary hunter who pursued it.

An unforgettable tale, magnificently told—an epic of beauty, terror, survival, and redemption for the ages.